BIOTECHNOLOGY FOR THE ENVIRONMENT
SOIL REMEDIATION
VOLUME 3B

FOCUS ON BIOTECHNOLOGY

Volume 3B

Series Editors
MARCEL HOFMAN
Centre for Veterinary and Agrochemical Research, Tervuren, Belgium

JOZEF ANNÉ
Rega Institute, University of Leuven, Belgium

Volume Editors
SPIROS N. AGATHOS
Université Catholique de Louvain,
Louvain-la-Neuve, Belgium

WALTER REINEKE
Bergische Universität,
Wuppertal, Germany

COLOPHON

Focus on Biotechnology is an open-ended series of reference volumes produced by Kluwer Academic Publishers BV in co-operation with the Branche Belge de la Société de Chimie Industrielle a.s.b.l.

The initiative has been taken in conjunction with the Ninth European Congress on Biotechnology. ECB9 has been supported by the Commission of the European Communities, the General Directorate for Technology, Research and Energy of the Wallonia Region, Belgium and J. Chabert, Minister for Economy of the Brussels Capital Region.

Biotechnology for the Environment: Soil Remediation Volume 3B

Edited by

SPIROS N. AGATHOS
Université Catholique de Louvain,
Louvain-la-Neuve, Belgium

and

WALTER REINEKE
Bergische Universität,
Wuppertal, Germany

KLUWER ACADEMIC PUBLISHERS
DORDRECHT / BOSTON / LONDON

A C.I.P. Catalogue record for this book is available from the Library of Congress.

ISBN 1-4020-1051-6

Published by Kluwer Academic Publishers,
P.O. Box 17, 3300 AA Dordrecht, The Netherlands.

Sold and distributed in North, Central and South America
by Kluwer Academic Publishers,
101 Philip Drive, Norwell, MA 02061, U.S.A.

In all other countries, sold and distributed
by Kluwer Academic Publishers,
P.O. Box 322, 3300 AH Dordrecht, The Netherlands.

Printed on acid-free paper

Printed in the Netherlands.

EDITORS PREFACE

At the dawn of the 21st century we are witnessing an expanding human population in quest of survival and continued well-being in harmony with the environment. Many segments of society are increasingly preoccupied with the battle against both diffuse and concentrated pollution, the remediation of contaminated sites, the restoration of damaged areas due to anthropogenic activities and the re-establishment of functioning biogeochemical cycles in vulnerable ecosystems. There is an enhanced awareness of the value of pollution prevention and waste minimization in industrial, urban and agricultural activities, as well as an increased emphasis on recycling. Faced with these major contemporary challenges, biotechnology is emerging as a key enabling technology, and, frequently, as the best available technology for sustainable environmental protection and stewardship.

Although the activities of microorganisms and their subcellular agents have been recognized, studied and harnessed already for many years in the environmental arena, there is a new dynamism in the in-depth understanding of the molecular mechanisms underlying the functioning of microorganisms and their communal interactions in natural and polluted ecosystems, as well as an undeniable expansion of practical applications in the form of the new industry of bioremediation. A number of distinct but increasingly overlapping disciplines, including molecular genetics, microbial physiology, microbial ecology, biochemistry, enzymology, physical and analytical chemistry, toxicology, civil, chemical and bioprocess engineering, are contributing to major insights into fundamental problems and are being translated into practical environmental solutions and novel economic opportunities.

The book set «Biotechnology for the Environment», based on a compilation of some of the outstanding presentations made at the 9th European Congress on Biotechnology (Brussels, Belgium, July 11-15, 1999) and enriched with newly updated thematic chapters, captures the vitality and promise of current advances in the field of environmental biotechnology and is charting emerging developments in the beginning of the new millennium. This second volume, subtitled 'Soil Remediation' offers a view on methodology in the area of bioremediation, illustrating both the diversity and the importance of the multidisciplinary approaches for years to come. After a general opening exemplifying the different approaches in bioremediation of contaminated soil in which the current practice and trends are described, life cycle assessment based software tools for soil remediation planning are compared to calculate the potential environmental burdens of bioremediation and to help soil remediation planners to estimate the overall environmental performance of different remediation concepts.

Two contributions offer state-of-the-art descriptions of *ex situ* clean up technologies, wherein slurry decontamination reactor processes are the central point of discussion. The contaminated site itself is used as integral reactor for the microbial degradation of contaminants when using *in situ* clean up techniques. The principles of the approach and the successful application of aerobic and anaerobic *in situ* processes in the field are clearly described, including "natural attenuation" processes, in the third part of this volume. Two approaches complete the picture on bioremediation, immobilisation of pesticides and humification of nitroaromatic compounds leading to drastically reduced bioavailability and detoxification of contaminants at the affected site. Finally, phytoremediation is presented as a viable bioremediation technique using living green plants to degrade, to contain or to render harmless contaminants of the environment, including recalcitrant organic compounds or heavy metals.

The Editors hope that the integration of the depth of scientific fundamentals with the breadth of current and future environmental applications of biotechnology so evident in these selected contributions will be of value to microbiologists, chemists, toxicologists, environmental scientists and engineers who are involved in the development, evaluation or implementation of biological treatment systems. Ultimately, a new generation of environmental scientists should take these lessons to heart so that new catalysts inspired from the biosphere can be designed for safe, eco- and energy-efficient manufacturing and environmental protection.

      Spiros N. Agathos   Walter Reineke

TABLE OF CONTENTS

EDITORS PREFACE v
Table of Contents 1
PART 1 Current practice and trends 5
  Biological soil treatment 7
    Jürgen Klein 7
    1. Introduction 7
    2. Fundamentals 8
    3. Necessary preliminary investigations 9
      3.1. Degradability of contaminants 9
      3.2. Bioavailability 9
      3.3. Adjustability of the biological and physico-chemical conditions for
      biological degradation in the soil 10
    4. Bioremediation techniques 13
      4.1. *Ex situ* processes 13
        4.1.1. Bio-heap treatment 13
        4.1.2. Reactor processes 14
      4.2. *In situ* methods 16
    5. Re-use of the soil 18
    6. Bioassays for soils 18
    7. Perspectives 19
      7.1. Development trends 19
      7.2. Future perspectives 20
    References 20
  Life cycle assessment in soil bioremediation planning 23
    S. Volkwein 23
    Summary 23
    1. Life cycle assessment in soil remediation planning 23
    2. Overview of case studies and projects 25
    3. Example using the environmental balancing software 25
    4. "Environmental balancing of soil remediation measures" method 26
    5. "REC" method 27
    6. "Environmental/economic evaluation and optimising of contaminated sites
    remediation" method 28
    7. General comparison of the three software tools 29
    8. Comparison of the life cycle assessment approaches 30
    9. Conclusions 31
    References 32
PART 2 *Ex situ* clean up technologies 35
  The DMT-BIODYN-process 37
    Christoph Sinder, Jürgen Klein and Frank Pfeifer 37
    Abstract 37

1. Introduction 37
2. Materials and methods 38
   2.1. The principle of the DMT-BIODYN process.............................38
   2.2. Laboratory facility...39
   2.3. Pilot plant 39
   2.4. Sapromat system 40
   2.5. Chemical analysis...41
   2.6. Process-engineering characterization 41
3. Results..41
   3.1. Results of process-engineering characterization 41
   3.2. Results of PAH-Degradation experiments 43
4. Discussion 46
Acknowledgements 48
References 48
The slurry decontamination process..51
   R.H. Kleijntjens, L. Kerkhof, A.J. Schutter, K.Ch.A.M. Luyben, J.F. De Kreuk
   and J. Janse 51
   Summary 51
   1. Introduction 52
      1.1.   Recycling of contaminated solid waste 52
      1.2. Characteristics of contaminated soil, sediments and sludges 53
   2. Classification of treatment technologies 54
      2.1. *In situ* remediation...54
      2.2. Constructed "natural" systems/simple technologies....................54
      2.3. *Ex situ* processing..55
   3. Bioreactors 56
      3.1. Slurry bioreactors 57
      3.2. Solid state bioreactors 58
   4. Configuration of *ex situ* bioprocesses...................................:.......59
      4.1. *Ex situ* slurry bioprocess...59
      4.2. Slurry Decontamination Process 60
      4.3. Microbial breakdown in the SDP 62
      4.4. SDP-improvements: froth flotation in the DITS-reactor 63
   5. Scale-up..65
      5.1. Process economics...65
      5.2. Extensive (low cost) treatment of the fines 66
      5.3. Environmental efficiency of the SDP.......................................67
   6. Conclusions...68
   Acknowledgement..69
   References 69
PART 3 *In situ* clean up technologies 71
In situ biological soil remediation techniques..73
   Peter Middeldorp, Alette Langenhoff, Jan Gerritse and Huub Rijnaarts...........73
   1. Introduction 73
   2. Source zone remediation techniques 75

2.1. Reductive dechlorination at the Rademarkt, Groningen, The Netherlands
.... 77
   2.1.1. Introduction 77
   2.1.2. Pilot study.. 77
   2.1.3. Results 77
2.2. Aerobic *in situ* bioremediation techniques 79
3. Active plume management techniques 79
   3.1. HCH removal at an industrial location in The Netherlands.................... 80
   3.2. Mineral oil and BTEX removal at a harbour site in The Netherlands 81
      3.2.1. Introduction 81
      3.2.2. Performance of the fences 81
      3.2.3. Technical comparison.. 81
   4. Natural attenuation 82
      4.1. Indicators for *in situ* biodegradation... 83
      4.2. Natural attenuation rates and plume behaviour 85
      4.3. Difficulties and complicating factors 86
      4.4. Plume management through monitored natural attenuation (MNA) 86
   5. Discussion and outlook 86
   References 89
PART 4 Immobilisation of pollutants in the soil.. 91
Immobilisation of pesticides in soil through enzymatic reactions.................... 93
   Jean-Marc Bollag 93
   Abstract 93
   1. Introduction 93
   2. Reactions between pesticides and humic material 94
      2.1. Covalent binding by soil microorganisms 94
      2.2. Oxidative coupling 95
   3. Enzymes and their origin... 95
      3.1. Peroxidases... 95
      3.2. Polyphenol oxidases 95
      3.3. Function of enzymes in binding reactions between pesticides and humic
      material... 96
   4. NMR spectroscopy to determine the type of binding of pesticides in the soil96
   5. Stability and release of bound pesticides... 98
   6. Enzymes as decontaminating agents 98
   7. Conclusions 99
   References 99
Humification of nitroaromatics 103
   Dirk Bruns-Nagel, Heike Knicker, Oliver Drzyzga, Eberhard von Löw,
   Klaus Steinbach, Diethard Gemsa 103
   1. Introduction 103
   2. Composting of soil contaminated with nitroaromatics............................... 103
   3. Optimisation of composting of TNT-contaminated soil............................. 104
   4. Transformation of TNT during anaerobic/aerobic composting.................. 106
   5. ^{14}C-TNT balancing in anaerobic/aerobic composting 107

6. Qualitative description of non-extractable ^{15}N-TNT residues formed by an anaerobic/aerobic composting 109

7. Conclusion 110

References 111

PART 5 Phytoremediation 113

  Phytoremediation 115

    Thomas Macek, Martina Mackova, Petra Kucerova, Ludmila Chroma, Jiri Burkhard and Katerina Demnerova 115

    Summary 115

    1. Introduction 115

      1.1. General introduction 115

      1.2. Phytoremediation and rhizoremediation 116

        1.2.1. Metabolic aspects of phytoremediation 116

        1.2.2. Rhizoremediation 117

        1.2.3. Exudates and enzymes released 118

      1.3. Methods used in phytoremediation 118

        1.3.1. Phytostabilisation 119

        1.3.2. Phytoextraction 119

        1.3.3. Rhizofiltration 119

        1.3.4. Phytodegradation 119

        1.3.5. Phytovolatilisation 120

        1.3.6. Hydraulic control 120

      1.4. Artificial wetlands 120

      1.5. Perspectives of plants in detoxification in CWD 121

    2. Examples of practical phytoremediation experiments 121

      2.1. Examples of organic pollutants and xenobiotics removal 122

      2.2. Examples of heavy metal removal 122

      2.3. Advantages and economical aspects 123

    3. Basic research aspects 124

      3.1. Plant *in vitro* cultures in phytoremediation studies 124

        3.1.2. Callus and cell suspension cultures 124

        3.1.3. Hairy root cultures 125

      3.2. Heavy metals 126

      3.3. Metabolism of PCBs by plants 127

        3.3.1. The enzyme systems involved 128

        3.3.2. PCB transformation products 128

        3.3.3. Pot and field experiments, rhizoremediation 129

        3.3.4. Improvement of plant material for PCB degradation 129

    4. Genetic modifications 129

      4.1. Breeding and genetic engineering 129

      4.2. Improvement of the degradation of organic compounds 130

      4.3. Improvement of heavy metal uptake 131

    5. Conclusions 132

    Acknowledgement 133

    References 133

INDEX 138

PART 1
CURRENT PRACTICE AND TRENDS

BIOLOGICAL SOIL TREATMENT

Status, development and perspectives

JÜRGEN KLEIN
Deutsche Montan Technologie GmbH
Division GUC Am Technologiepark 1, D-45307 Essen, Germany.
E-mail: Sinder@dmt.de

1. Introduction

Environmental protection over the past few decades has meant primarily the protection of air and water. Only with the increasing use of land in industrialised societies and the highlighting of possible hazards from contaminated soil did the public become aware of soil protection in the early 80s. This has also prompted industry to take up this market segment. Engineers and scientists have thus been spurred on to look for technically optimised, ecologically sound and economically appropriate solutions.

For more than a century natural biochemical processes (nature's self-cleaning forces) have been utilised to treat effluent, and reactors and plant systems had been adapted with increasing effect to cope with the difficult conditions, but only in the 80s was work begun on testing biological methods for cleaning up soil. The experience accumulated in biological soil clean up in the first few years was characterised both by successes and failures, including those of unprofessional suppliers. Initially therefore the acceptance of biological soil clean up was only limited. But now, thanks to intensive and interdisciplinary work, impressive successes are in evidence. Consequently biological soil clean-up methods now enjoy a high technical level and a broad acceptance (Klein 1996).

In 1998 approximately 2.2 million tonnes of soil were remediated in Germany in 108 stationary soil treatment facilities, of which 1.2 million tonnes or 60 % was treated by biological means. For on-site biological treatment there is an available capacity of about 0.4 million t/a (Table 1).

The aim is to maintain and apply this level, even if the prospects for soil clean up in Germany are seen from a more modest point of view. In addition to the preference for securing techniques and to the competition from less expensive suppliers in neighbouring countries, low-price dumping has in recent times represented a challenge to operators of soil decontamination facilities in Germany.

S.N. Agathos and W. Reineke (eds.).
Biotechnology for the Environment: Soil Remediation, 7-21.
© 2002 Kluwer Academic Publishers. Printed in the Netherlands.

Table 1: Soil treatment plants (status December 1997)

Technology used		Biological	Scrubbing	Thermal	Total
Treatment Centres	Number	80	24	4	108
Capacity	in Mio.t/a	1.9	1.38	0.17	3.45
	in %	55	40	5	100
Throughput	in Mio. t/a	1.24	0.85	0.11	2.20
	in %	56.4	38.7	4.9	100
Mobile Plants:	Number	8	7	5	20
Capacity	in Mio.t/a				~1.0
Contaminants handled		TPH	TPH	TPH	
		BTEX	BTEX	BTEX	
		Phenol	Phenol	Phenol	
		PAH	PAH	PAH	
		VOCH	VOCH	VOCH	
			Pesticides	Pesticides	
			PCB	PCB	
			PCDD/F	PCDD/F	
			Cyanide	Cyanide	
			Phosphor		
				Heavy metals	Heavy metals

TPH: total petroleum hydrocarbons, BTEX: benzene, toluene, ethylbenzene, xylene; PAH: polycyclic aromatic hydrocarbons; VOCH: volatile halogenated hydrocarbons; PCB: polychlorinated biphenyls; PCDD: polychlorinated dibenzodioxin; PCDF: polychlorinated dibenzofurans

2. Fundamentals

The use of biological processes for treatment of liquid wastes from human activities is an established technology dating back at least 4000 years. The knowledge of the biodegradation mechanisms of organic pollutants and especially of synthetic compounds, however, is more recent and has only been developed in the second part of the 20th century.

In biological processes, use is made of the capacity of microorganisms to consume organic substances as a nutrient (substrate) and to convert them to harmless natural materials, such as CO_2, water and biomass. For degradation of noxious substances in contaminated soils, bacteria and fungi are of foremost importance.

The attack of organic substances in soils by microorganisms results either in full degradation (mineralisation) or in a partial degradation process producing metabolites which may be used by other members of the biocenosis or which remain in the soil. Furthermore, the original substance and the metabolites can be cycled to the soil's carbon depot. This is called humification. The decisive point in that case is that the noxious matter or the resulting metabolite is incorporated into the soil matrix, which reduces sharply their availability for biochemical reactions. This phenomenon is used at present in comprehensive development work aimed at a conversion of the pollutants by microbiological attacks so they are converted to natural substances, such as humic compounds, which involve no further environmental risk. Detoxification can also be obtained by co metabolism using non-growth substances as co substrate.

3. Necessary preliminary investigations

The decision in favour of a biological decontamination process depends on the following prerequisites:

- Degradability of the contaminants

- Bioavailability of contaminants in the soil matrix

- Adjustability of the biological, physical and chemical conditions required for biological degradation in the soil.

3.1. DEGRADABILITY OF CONTAMINANTS

Laboratory methods for the evaluation of biological soil cleanup processes have been developed by a *Dechema Working Group* (DECHEMA 1992) allowing findingout thoroughly about the microbial degradability of contaminants. However, it has to be considered that the assessment of degradability of soil contaminants requires investigation on the original soil, either in suspension or by means of naturally moist samples.

3.2. BIOAVAILABILITY

In many cases it is not the actual microbial degradability but physico-chemical parameters, such as adsorption/desorption, diffusion, and solution properties of the contaminants found in the soil, frequently in solid phase, which are decisive for degradation and the degradation rate. The availability of contaminants in the soil to the microorganisms, the bioavailability, is decisively influenced by the configuration of the soil matrix, i.e. its material composition and particle size distribution, and also by the history of contamination. For these cases, suitable bioavailability assessment methods are on hand (ITVA 1994).

Figure 1 shows a sequence pattern for close-to-practice preliminary investigations. When following that scheme we arrive - with relatively low expenditure - first at the important decision whether biological decontamination is possible at all. If there is no bioavailability of contaminants in the soil, or if a contamination, which is toxic to microorganisms, cannot be eliminated, this means that the soil concerned is out of question for a microbiological treatment.

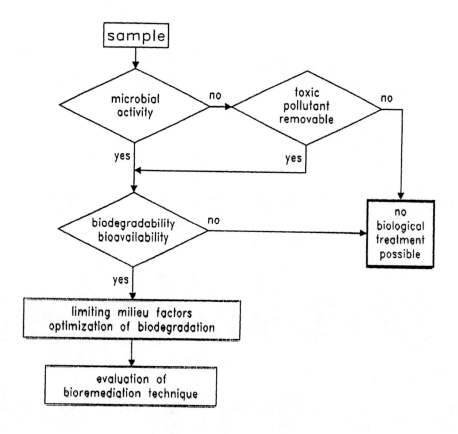

Figure 1: Evaluation of the biodegradability of a pollutant

3.3. ADJUSTABILITY OF THE BIOLOGICAL AND PHYSICO-CHEMICAL CONDITIONS FOR BIOLOGICAL DEGRADATION IN THE SOIL

The physico-chemical and of course, also the geological properties of the soil are decisive for the choice of methods for bioremediation.

Figure 2 shows the sequence of investigations in view of the selection of methods. The decision in favour of an *in situ* or an *ex situ* method generally depends on the hydro-geological configuration of the soil, the permeability coefficient k_f, the soil's homogeneity, and its silt and fines content. The k_f value is regarded as orientation parameter. Experience has shown that with k_f values $< 10^{-5}$ m/s, *in situ* treatment is out of question. Only in relatively few cases, favourable conditions for a microbiological *in situ* decontamination prevail.

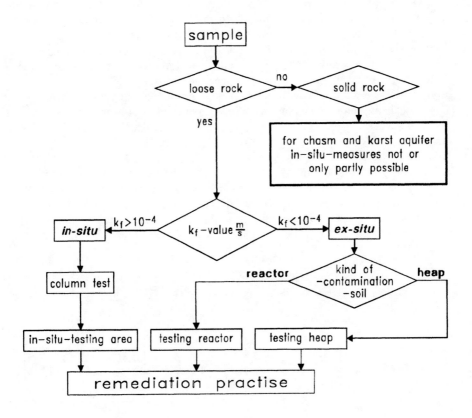

Figure 2: Evaluation of bioremediation techniques

Table 2 contains a list of the substance groups, which count as relevant in terms of contaminated sites, together with an assessment of their microbial degradability (DECHEMA 1991; ITVA 1994). It is not possible to define in general terms a concentration range within which microbial degradation is possible, but rather this differs according to the substance class. Similarly it is not possible to fix a generally applicable value for an achievable residual concentration, since this depends on the type and starting concentration of the pollutants and the adjustable ambient conditions.

To date the soils most frequently cleaned have been those contaminated with oil. The degradability of aliphatic hydrocarbons, such as petrol, diesel and other oil derivates, can be described as good.

The persistence increases with the chain length and the degree of branching, and so for compounds with more than 25 carbon atoms the degradation rate slows down, if it does not come to a complete standstill.

It is more difficult to clean soils from the coal by-product domain (gas works, coking plants), where the contamination is mainly in the form of BTEX aromatics (benzene, toluene, ethylbenzene, xylenes) and polycyclic aromatic hydrocarbons (PAH).

Whereas 4-ring PAHs can now be described as being highly degradable, it is not always possible to degrade PAH with 5 and more aromatic rings. If degradation is encountered here, then this is due to co metabolism, which means that the microorganisms do not use these aromatics as the sole source of carbon and energy, but can only use them jointly in the presence of other hydrocarbons (cosubstrates).

In many cases success is experienced with the biological treatment of soils contaminated by highly volatile, halogenated hydrocarbons (VOCH).

In contrast the degradation of polychlorinated biphenyls (PCB) and polychlorinated dibenzodioxins and dibenzofurans (PCDD/PCDF) should be regarded in principle as very difficult. To date it has not been possible to provide a scientific verification of the complete dehalogenation of polychlorinated dioxins.

Heavy metals are in principle not degradable by microbial means. Under certain conditions they can be adsorbed on biomass by biosorption or taken up by plants from the soil.

Table 2: Biodegradability of soil contaminants

Class of Contaminants	Basically well degradable	Basically hard degradable
aliphatic hydrocarbons (HC), petrol-HC and derivates	+	
monocyclic aromatic (e.g. BTEX) and heterocyclic (e.g. Pyridin, Chinolin) HC	+	
polycyclic aromatic HC (PAH)	+[1]	+[2]
volatile halogenated, especially chlorinated hydrocarbons (VOCH)	+	
alicyclic VOCH and derivatives (e.g. HCH)	+	
polychlorinated biphenyls (PCB)		+*
polychlorinated dibenzodioxins and dibenzofurans (PCDD resp. PCDF)		+*
pesticides and derivatives		+*
heavy metals	not degradable	

*some low chlorinated congeners are principally degradable/dehalogenable, degradation of highly chlorinated congeners cannot be demonstrated at present [1] up to 4-Ring-PAH [2] 5- and 6-Ring-PAH

4. Bioremediation techniques

Figure 3 shows the generally practicable process options. After the required preliminary investigations and balancing of the microbial degradation processes to the greatest extent possible, a decision in favour of an *ex situ* or *in situ* method is made.

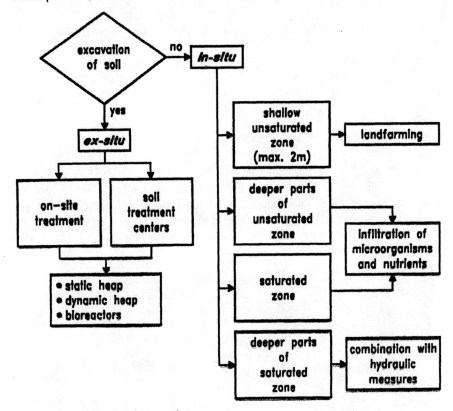

Figure 3: Bioremediation techniques

4.1. *EX SITU* PROCESSES

These processes require excavation of the contaminated soil masses and their treatment either in bio-heaps or in reactors, either on-site or off-site in a soil decontamination plant, which requires transport capacity.

4.1.1. Bio-heap treatment

As to heap treatments, difference is made between static and dynamic ones. In static bio-heap methods, irrigation and aeration networks are installed for ensuring the necessary environment conditions.

In the case of dynamic heap methods the deposited soil masses are, in specified intervals, turned over, homogenized, and spread. During these turning-over cycles, further water and nutrients are added, if necessary.

13

4.1.2. Reactor processes

Reactor techniques are made use of for shortening the time for the treatment of soils with high fines portions and high rates of contamination as well as for better process control with respect to exhaust air and wastewater treatment. According to the water content of the soil, difference can be made between solid state and slurry state methods (Mann *et al.* 1995).

Solid-state processes. In solid-state processes, the soil is treated at a humidity rate corresponding approx. to 50 to 70 % of its maximum water capacity. The treatment is run in rotary reactors or in static reactors equipped with internal mixing systems. The principle of these methods is largely identical to the one of the dynamic heap processes. In both cases the biological degradation of noxious substances contained in the soil is arrived at by agitation and aeration of the naturally moist soil. Also the treatment of soil/ compost blends is to be categorized as solid-state process. These dry processes are advantageous in that the treated soil does not need to be dewatered.

A reactor, which is mainly used for mobile application on-site, is shown in Figure 4. The reactor allows the treatment of various amounts of soil by connecting together the appropriate number of container-like reactors to the necessary length respectively reactor volume (each reactor with a dimension of 5 × 1 m). A horizontal driven mixing device realizes a sufficient mixing and a homogeneous distribution of nutrients in the soil. The system is operated automatically controlled and allows either aerobic or anaerobic treatment as well as alternating combinations.

Figure 4: Example for a mobile solid-state bioreactor (Courtesy Umweltschutz Nord)

Slurry state processes. Extremely fine-grained or cohesive soils, or sludge, e.g. products from pre-treatment steps such as soil washing, cannot be purified by heap or solid-state reactor techniques. Accordingly, slurry state techniques are run for treatment of those soils. The soil portion per weight of slurries ranges between 30 and 50 % by weight. Processes run in impeller-type mixing vessels, airlift or fluidised-bed reactors, which provide sufficient and homogeneous nutrient and oxygen supply, are particularly suitable.

The key criteria for these processes read:

- The particle sizes are less than 1000 μm.

- All soil particles must be in full contact with the aqueous phase.

- In particular for PAH, long residence times are to be catered for.

- Due to the long residence times the energy requirement must be low and the solids portion of a reactor fill must be high.

As an example a process developed at Delft University of Technology and consisting out of a series of tapered aerated slurry bioreactors is shown in Figure 5.

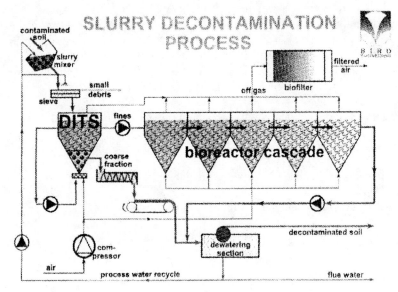

Figure 5: Example for a slurry-state bioreactor (SDP Process)

The SDP-process consists of six steps (Kleijntjens 1999):

- Mixing of soil with water to produce the slurry.

- Wet sieving of the slurry, the coarse fraction normally not contaminated is being separated.

- Treatment in the Dual Injected Turbulent Separation (DITS-) reactor, a combination of separation and decontamination. The coarse soil fraction can be separated, washed and removed to this first reactor. The fines are fed into the ISB-cascade.

- Further treatment in the Interconnected Suspension Bioreactor (ISB-) Cascade. Here the less available part of pollutants is degraded.

- In the dewatering step the outgoing flow of treated soil are pressed to a filter cake or dewatered by other means.

- Off-gas treatment by activated carbon filter or biofilter.

At the present time, a prototype plant with a reactor volume of 700 m³ to process 15.000 tonnes a year (8-10 days residence time) is in operation.

Another fluidised bed process has been developed by DMT, Essen (Sinder 1999). The fine-grained material is brought into suspension in a hydraulically run fluidised bed reactor. Due to the necessary long residence times, e.g. for PAH degradation, the process is run batch wise. The larger cross section in the reactor's top zone assures the sizing effect and a reduction of the solid's concentration in the slurry. Besides external aeration, also an internal aeration can be activated on option to cater for increased oxygen consumption. The inflow velocity relative to the reactor cross-section in the centre zone of the reactor is of roundabout 10^{-3} m/s. This allows suspension of an approx. 4 to 8 m high bed and a solids concentration of approx. 50 % by weight. These data also allow corresponding scaling up. A commercial plant with a reactor volume of 400 m³ is in operation in Sweden to clean PAH-contaminated soil.

The advantages of the application of bioreactors summarized can be read:

- Decontamination takes place under well-controlled conditions easy to optimize

- Treatment of all kinds of soils even with high ratios of fines

- Treatment of soil contaminated with hazardous substances

- Aerobic and anaerobic as well as alternating aerobic / anaerobic treatment possible

- Increasing of bioavailability especially in slurry reactors

- Optimum emission control

4.2. *IN SITU* METHODS

The decisive viewpoint for an *in situ* treatment is that no soil masses need to be removed and that the saturated or unsaturated zone of a contaminated site itself is used as integral reactor for the microbial degradation of contaminants. Figure 6 shows the typical features of an old pollution. The contaminants enter the soil and migrate downwards. The amount of contaminants remaining in the unsaturated zone is controlled by sorption and diffusion into the soil pores and retention by capillary forces. Generally the contaminants may form - when they reach the groundwater table - a separate phase on top of the groundwater so-called NAPLs (non-aqueous phase liquid) or at the basis of the aquifer (groundwater zone) DNAPLs (dense non-aqueous phase liquid). Minor amounts of the contaminants are dissolved in the groundwater and are transported with the natural groundwater flow forming the contaminated plume. The spatial extent of the plume and the level of the contaminant concentrations depend on the age of the pollution, the sorption and transport characteristics as well on the efficiency of the natural degradation. Depending on the features of the contamination and of the site different technologies may be chosen. The technologies are subdivided into technologies for the treatment of the unsaturated and the saturated zone. A further subdivision comprises highly active technologies (bioventing, bioslurping, biosparging and hydraulic circuits) as well as active and passive technologies (funnel and gate, bioscreen and natural attenuation)

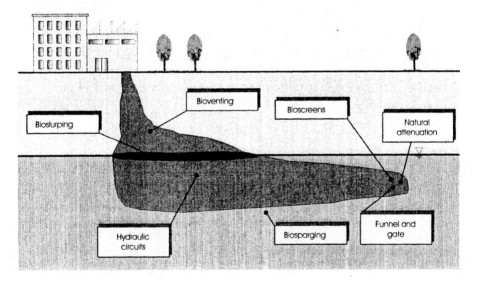

Figure 6: Bioremediation alternatives

For shallow unsaturated zones, land farming is possible, which allows addition of nutrients and the adjustment of the environment conditions. In deeper unsaturated zones and in parts of the saturated zones, infiltration of microorganisms (bioaugmentation) and nutrients by a flushing circuit, sometimes combined with hydraulic measures is practised. The availability of molecular oxygen - dissolved in water and in this way cycled to the process - is decisive for rapid aerobic degradation. In such a process configuration, the addition of an oxygen carrier (such as hydrogen peroxide or industry grade oxygen) may be advantageous.

For decontamination under anoxic or anaerobic conditions (e.g. in the unsaturated zone) another electron acceptor (such as nitrate) is necessary to be used for the microbiological metabolism.

The main features of biological soil remediation are summarized in Table 3.

Table 3: Comparison of ex situ and in situ bioremediation techniques

Characteristics	Treatment	
	ex situ	*in situ*
Removal of bioavailable contaminants by mineralisation to CO_2 and H_2O	+	+
Use of indigenous microorganisms	+	+
Pre-investigations needed for evaluation of biodegradability and remediation-techniques	+	+
Operation time depends on kind and concentration of contaminants	+	+
Prediction of remediation success is difficult		+
Treatment of large amounts of soil possible,		
if suitable areas are available	+	
if there are no remediation time limits		+
Flexible adaptation to site conditions e.g. remediation under buildings or on built-up areas		+
Low energy demand	+	+
Control of degradation conditions required during the total remediation period	+	+
Suitable for ad hoc measures, e.g. accidents		+

5. Re-use of the soil

A major aspect of biological soil remediation is the re-use of the treated soil. Structural use includes, for example, backfilling noise baffle embankments, roadside fortification or ditches and landfill measures. The vegetation-related use ranges from roof greening and parks to agricultural use. The latter requires prior testing on large areas with grass as the test seed. Already a number of positive results have been obtained where it is possible to recultivate corn and potatoes after such a test.

6. Bioassays for soils

A crucial factor for the re-use of soils is a toxicological/ecotoxicological assessment. For this purposes use is made of bioassays to measure the impact of contaminated or treated soils. Biological tests have proven especially beneficial if chemical or environmental samples of complex composition have to be tested with respect to their hazard potential. They integrate the effects of all acting substances, including those not considered or recorded in chemical analysis. A *Dechema Working Group* (DECHEMA 1995) has reported the state of the art of scientific development and experience and has recommended an assessment strategy which meets practical requirements:

- Preliminary assessment of contaminated sites. Recognition of ecotoxicological effects, caused by contaminants, which might not have been detected by chemical analysis.

- Ecotoxicological process control during biological soil clean up.

- Evaluation of a possibly residual toxic/ecotoxic potential of treated soils before their further use.

7. Perspectives

The euphoria that has accompanied soil decontamination in the last few years has now given way to a realistic attitude of more appropriate proportions. The discussion on the equivalence of securing and decontamination techniques must not lead to a clean up on a low level, but rather to ecologically and economically appropriate solutions.

The original objective of multifunctional use through the restoration of a "natural" soil is not feasible in most cases for technical and financial reasons. In view of this biological clean-up techniques still have to be improved and optimised in order to provide cost-effective, technically simple and near-natural processes.

7.1. DEVELOPMENT TRENDS

Solutions attempted for *ex situ* processes, for example, are the combination of fixed-bed and liquid-phase bioreactors, which provide a great degree of flexibility in adapting the required ambient conditions to the relevant contaminant and also the special features of the soil matrix, or for *in situ* measures the dosing of tensides and heating of the soil by radio waves, for example, by which means it is possible to enhance the biological availability and hence the degree of degradation (Stottmeister 1998).

The treatment principle of humification is currently the subject to considerable attention in a joint BMBF project concerned with the clean-up of contaminated armaments sites, primarily the contaminant TNT. To eliminate TNT the only reasonable approach is co metabolic transformation with the subsequent binding of the products arising in the soil matrix. Both certain bacteria and fungi are capable of the co metabolic transformation of TNT to reduced metabolic products, such as aminonitrotoluenes. These are then partly degraded, mineralised or bound to the soil matrix in a process similar to natural humification. Both *ex situ* and *in situ* technologies are being studied.

For example, anaerobic/aerobic combination processes (modified composting) are being tested as an *ex situ* technique (Umweltschutz-Nord). In the development of suitable *in situ* methods, an investigation is being conducted into whether an accumulation of TNT in the rhizosphere of plants and undergrowth can be economically exploited. This method, known as phytoremediation, is being researched mainly in the USA, Britain and Germany with a view to the further development of near natural, *in situ* clean-up processes for the accumulation of metals and organic compounds

Successful research has also been conducted into the activation of natural biocenoses in the clean-up of sediments and acidic waters from uranium mining, as is demonstrated by the three-year-long operation of a pilot plant for heavy metal leaching (Stottmeister 1998).

Since all biological clean-up processes are based on nature's own self-purification forces, they are the subjects of increasing attention and methods are being propagated under the names of "intrinsic bioremediation" or "natural attenuation" (Klein 1998).

The USEPA policy directive defines monitored natural attenuation as "the reliance of natural attenuation processes (within the context of a carefully controlled and monitored site cleanup approach) to achieve site-specific remediation objectives within a time frame that is reasonable compared to that offered by more active methods"

In neither case are the terms used to mean that nothing is done and everything is left up to nature, which would not be acceptable anyway in the case of an actual hazard. Rather they involve more research work than is necessary for an engineering solution. The solution to the following questions is implied:

How can suitable microorganisms most effectively be incorporated and distributed in the soil (bioaugmentation) and stimulated to optimum activity (biostimulation) and how must the groundwater be guided in order to render an in situ treatment possible and effective? With regard to the latter, in addition to the common techniques for groundwater treatment (pump and treat), techniques propagated in the USA under the names funnel and gate systems and reactive barriers are being tested increasingly. The most important restriction of such near-natural processes is the time span available to clean up the site.

7.2. FUTURE PERSPECTIVES

Suitable and sustainable solutions require a greater interdisciplinary approach, in other words the involvement of microbiologists, geneticists, chemists, hydrogeologists, soil scientists and process engineers (Knackmuss 1998).

A suitable and sustainable strategy for avoiding environmental contamination can also only be achieved by an interdisciplinary collaboration between all protagonists in research and industry. The wide-ranging experience accumulated with respect to the contamination of soils and groundwater must be provided special impetus for testing the environmental impact of new chemical products before they are introduced, thus preventing subsequent contamination. A benign by design chemistry would therefore have to concentrate its research on identifying forms of bonding which facilitate the development of biodegradable and environmentally sound substances in the circulation of chemical products.

References

DECHEMA e. V. (1991) Einsatzmöglichkeiten und Grenzen der mikrobiologischen Verfahren zur Bodensanierung DECHEMA e. V., Frankfurt, June 1991.

DECHEMA e. V. (1992) Laboratory Methods for the Evaluation of Biological Soil Clean-up Processes DECHEMA e. V., Frankfurt, June 1992, ISBN 3-926959-39-X.

DECHEMA e. V (1995) Bioassays for Soils DECHEMA e. V., Frankfurt, 1995, ISBN 3-926959-52-5 In situ-Sanierung von Böden 11. DECHEMA-Fachgespräch Umweltschutz / ISBN 3- 926959-72-X 17./18. 04.1996, Frankfurt am Main.

Held T. and Dörr H. (2000) In situ RemediationBiotechnology, 2nd Edition, Vol 11b, Eds.: H.-J. Rehm, G. Reed, A. Pühler, P. Stadler. Wiley-VCH, p. 350-370.

ITVA (1994) Mikrobiologische Verfahren zur Bodendekontamination ITVA-Arbeitshilfe 1994 Hrsg.: Ingenieurtechnischer Verband Altlasten e.V. ITVA Berlin.

Kleijntjens, R. (1999) The Slurry Decontamination Process: Bioprocessing of contaminated solid waste streams, Proceedings ECB9, Brussels, 1999.

Klein, J. (1996) Möglichkeiten und Grenzen der biotechnologischen Bodensanierung FAZ-Beilage Biotechnologie 24.09.1996.

Klein, J. (1998) Quo vadis biotechnologische Bodensanierung FAZ-Beilage Biotechnologie 13.10.1998.

Knackmuss, H.-J. et al. (1998) The Take Home Message, Innovative Potential of Advanced Biological Systems for Remediation Workshop March 2-4, 1998, Technical University Hamburg-Harburg.

Mann, V., Klein, J.; Pfeifer, F.; Sinder, C.; Nitschke, V.; Hempel, D. C. (1995) Bioreaktorverfahren zur Reinigung feinkörniger, mit PAK kontaminierter Böden TerraTech 3 (1995), S. 69-72.

Sinder, Ch. The DMT-BIODYN-Process: a suspension reactor for biological treatment of fine grained soil, Proceedings ECB9, Brussels, 1999.

Stottmeister, U. (1998) Trends in der Entwicklung der biologischen Methoden in der Sanierungsforschung, Vom Wasser, 91, 343-350 (1998).

LIFE CYCLE ASSESSMENT IN SOIL BIOREMEDIATION PLANNING

S. VOLKWEIN
C. A. U. GmbH, Daimler Str. 23, D-63303 Dreieich, Germany
Phone +49 - 61 03 - 98 30 FAX +49 - 61 03 - 9 83 10
Email: s.volkwein@cau-online.de

Summary

In several cases, life cycle assessment has been used as one supplementary tool in the planning of soil remediation. Three life cycle assessment based software tools for soil remediation planning are compared. One example of soil treatment in a bioreactor is investigated by using the software "Umweltbilanzierung von Altlastensanierungs-verfahren" (version 1999).

The decision support tool "risk reduction – environmental merit – costs" (REC or RMK, internet version released in 2000) combines risk analysis, cost analysis with an approach, which is similar to life cycle assessment. The software "Umweltbilanzierung von Altlastensanierungsverfahren" ("Environmental balancing of soil remediation measures") of the German state Baden-Württemberg compares the environmental burdens of the remediation with the results of a risk assessment. The software "Environmental/economic evaluation and the optimising of contaminated site remediation" analyses "environmental costs" and "Environmental benefits" with an life cycle assessment approach.

The main innovative part in the three software products is the calculation of potential environmental burdens (Screening life cycle assessments) of soil remediation measures.

The degree of transparency of the three different life cycle assessment approaches is discussed.

1. Life cycle assessment in soil remediation planning

Remediation of contaminated sites is done to improve the environment. This improvement can be diminished by the possible environmental impacts of the technical measures. The planning of soil remediation concepts requires often the use of many tools. Such tools can be financial assessments, legal assessments or risk assessments. All useful tools together form a toolbox for the remediation of contaminated sites. This toolbox should include tools to evaluate the environmental burdens caused by the

23

S.N. Agathos and W. Reineke (eds.).
Biotechnology for the Environment: Soil Remediation, 23-33.
© 2002 *Kluwer Academic Publishers. Printed in the Netherlands.*

remediation (itself). Several governmental organisations promote the use of such tools (for example Volkwein 2000a). In the Netherlands (REC method), in Denmark (Environmental/economic evaluation and optimising of contaminated sites remediation) and in Germany (Environmental balancing of soil remediation measures), software tools have been developed for the consideration of the environmental burdens of the remediation (itself).

The evaluation of the overall environmental performance of a specific soil remediation option requires the knowledge of the environmental burdens caused by the remediation (itself) as discussed by Volkwein (2000c). The life cycle assessment (LCA) method is a suitable tool to assess the potential environmental impacts of remediation. The life cycle assessment method is still under development, but practitioners can refer to four international standards (ISO 14040:1997, ISO 14041:1998, ISO 14042:2000, ISO 14043:2000). Due to the data uncertainty in the soil remediation-planning phase the requirements for the use of life cycle inventory data are relatively low.

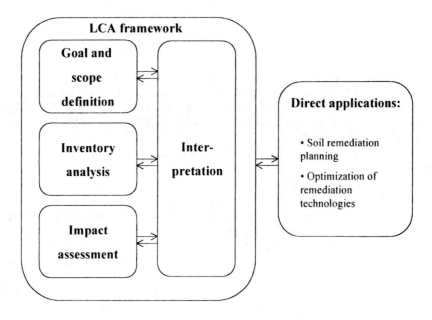

Figure 1: Life cycle assessment (LCA) framework adapted from ISO 14040:1997 with special direct applications for soil remediation planning

According to the ISO standards, every life cycle assessment consists of four parts:(as shown on Figure 1):
- goal and scope definition
- life cycle inventory
- life cycle impact assessment
- life cycle interpretation

2. Overview of case studies and projects

Joziasse *et al.* (1998) and Diamond *et al.* (1999) apply life cycle assessment to the remediation of contaminated sites. These studies are restricted to selected remediation cases.

Beinat *et al.* (1997) developed a decision tool (REC) integrating many options in making remediation. In 2000, the software tool RMK (English REC) in Dutch language has been published on the website of the Vrije Universiteit Amsterdam (van Drunen *et al.* 2000).

In 1999, the Landesanstalt für Umweltschutz Baden-Württemberg published a software tool in German language called "Umweltbilanzierung von Altlastensanierungsverfahren" (Landesanstalt für Umweltschutz Baden-Württemberg 1999). The English name is "environmental balancing of soil remediation measures". The English description of the tool is given by Volkwein *et al.* (1999).

Lieber *et al.* (2000) describe the CIASR method (Comprehensive impact assessment of site remediation). This method combines the approach of the environmental balancing with a financial assessment and a social assessment.

In 2000, the "environmental/economic evaluation and optimising of contaminated sites remediation" software (Life, 2000) has been released. The software part is mainly based on life cycle assessment. The economic evaluation is not part of the software.

Handling data from detailed risk assessments are called in this paper risk assessment too. The relative importance of the risk assessment part in the three software increases from the environmental balancing over the environmental/economic evaluation to the REC method. The risk assessment approaches and cost approaches of the three software tools are outside of the scope of this paper.

Preliminary descriptions and case studies of the environmental balancing method have recently been published (Volkwein *et al.* 1997, 1998; Bender *et al.* 1998, 1999; Njiboer and Schelwald-van der Kley 1998).

3. Example using the environmental balancing software

One example of a life cycle assessment is described by Volkwein and Klöpffer (1999). The functional unit is the treatment of 100 Megagram contaminated soil in a bioreactor (TERRANOX, Umweltschutz Nord, Germany). The treatment of contaminated soil (Doetsch and Dreschmann 1992) is a fully automatic and closed bioreactor system. The system consists of a variable number of reactors. The soil in the reactor is continuously mixed. The residence time is between 36 hours and 30 days. Every soil treatment consists of an anaerobic and an aerobic phase. After the bioreactor treatment the soil is further treated in a heap (TERRAFERM process, Umweltschutz Nord). The temperature in the bioreactor has to be kept above ten degrees Celsius. The average electricity consumption is 15 kilowatt for 100 Mg during 10 days.

The modelling of the electricity mix in Germany and of other basic assumptions for the life cycle inventory analysis are described in Volkwein *et al.* (1999) and references cited therein. The electricity production in nuclear, coal, gas, oil and water power stations is taken from Frischknecht *et al.* (1995). Diesel consumption in machines, nutrients as sodium nitrate (made from sodium hydroxide and nitric acid) are calculated from data of Frischknecht *et al.* (1995).

The results of the life cycle inventory (output data) are listed in Table 1. The cumulative (primary) energy demand is differentiated in renewable, nuclear and fossil energy.

The life cycle impact assessment aggregates the results of the life cycle inventory analysis. Impact categories like global warming and acidification are used. The basic assumptions for the life cycle impact assessment are explained in Volkwein *et al.* (1997, 1998, 1999) and Bender *et al.* (1998). Table 1 presents a survey of the some of the parameters of the analysis.

Table 1: Life cycle interpretation parameters of soil treatment in a bioreactor

Impact categories and energy and waste	Unit	Amount
Cumulative energy demand	Gigajoule	52.3
Waste total	Megagram	100.4
Waste from contaminated site to landfill	Megagram	100
Fossil resources	kg/year	8.91
Water	m^3	135
Land use	m^2 year	122
Global warming (mainly CO_2)	kg CO_2 equivalents	3.170
Acidification (mainly SO_2 and NO_x)	kg SO_2 equivalents	7.96
Photo-oxidant formation (mainly NMVOC)	kg ethene equivalents	0.40
Toxicity air (mainly HF, NO_x, and particles)	$10^6 m^3$	461
Toxicity water (mainly Al and organics)	$10^3 m^3$	12.7
Toxicity soil (mineral oil)	kg	2.53
Odour (mainly SO_2 and NO_x)	$10^6 m^3$	45.7

NMVOC non methane volatile organic compound

The electricity consumption causes 96 % of the result for the global warming, human toxicity soil, and 95 % of the cumulative energy demand. If the goal of the life cycle assessment is the optimisation of the bioreactor process (improvement assessment) one has to look for possibilities in reducing the electricity consumption. If the goal of the life cycle assessment is the selection of the most suitable bioreactor the life cycle inventory tables for the different bioreactors have to be calculated. If the goal of the life cycle assessment is the comparison of the soil treatment in a bioreactor with the soil treatment with other techniques the appropriate life cycle inventory analysis has to be performed. In each application case of the life cycle assessment a specific sensitivity analysis has to be made. An example for linking life cycle assessment with risk assessment is shown by Volkwein *et al.* (1999).

4. "Environmental balancing of soil remediation measures" method

The language of the environmental balancing ("Umweltbilanzierung von Altlastensanierungsverfahren") software is German (Landesanstalt für Umweltschutz Baden-Württemberg 1999). The method is also called "Environmental balancing of soil remediation" or simply "Environmental balancing". The software calculates an life cycle inventory. In the life cycle impact assessment, the life cycle inventory is transformed into indicator results. A rough data quality analysis is performed. In the life cycle interpretation, disadvantage factors for selected life cycle inventory entries and impact indicator results (Volkwein *et al.* 1999) are calculated. A preliminary ranking of two soil remediation options compared can be made by the software user based on the table of disadvantage factors. Outside of the life cycle assessment framework, guidance for the improvement of remediation options is given. Also the table of disadvantage factors can be compared with the results of a separate risk assessment (area, volume) of contaminated soil and groundwater of the site before and after the remediation (Figure 2).

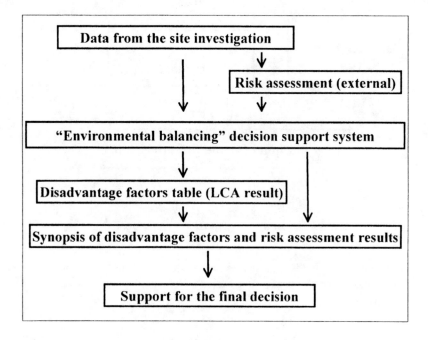

Figure 2: "Environmental balancing of soil remediation measures" method

5. "REC" method

Beinat *et al.* (1997) developed the decision support tool REC. REC is based on assessments of Risk reduction, Environmental merit and Costs. REC can help in comparing soil remediation options. The Dutch name is "RMK: een methodiek voor het vergelijken van bodemsaneringsvarianten op basis van de aspecten Risicoreductie, Milieuverdienste en Kosten". A more detailed description in English of the REC method is given by Nijboer and Schelwald-van der Kley (1998).

The REC method provides three indices (risk reduction, environmental merit and costs), (Figure 3).

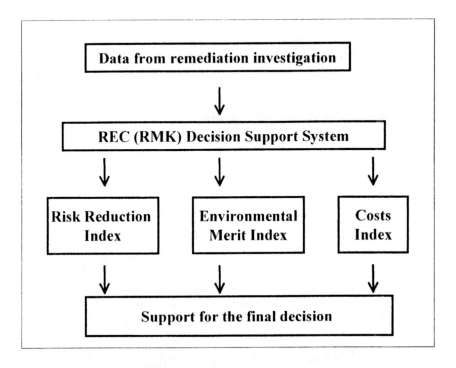

Figure 3: REC method in a decision making process

The environmental merit index is based on nine indices. Three of these indices (energy demand, air emission score, waste) are derived from a life cycle assessment of the remediation measures.

6. "Environmental/economic evaluation and optimising of contaminated sites remediation" method

A consortium of companies lead by the company ScanRail Consult developed the life cycle assessment based software tool "environmental/economic evaluation and optimising of contaminated sites remediation". The Excel ® based software is available in Danish and English. The software does not include the economic evaluation.

The life cycle assessment method is applied in two ways: the calculation of "environmental costs" and "environmental benefits". The benefits are mainly expressed in terms of life cycle indicator results of the life cycle impact categories human toxicity and ecotoxicity. Some data can be estimated as qualitative data. The single results are not aggregated to one overall score. Therefore, a synopsis is the key decision making process (Figure 4).

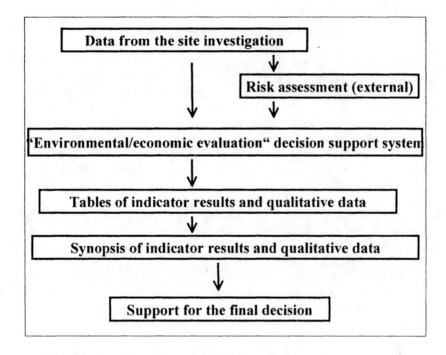

Figure 4: Software: Environmental/economic evaluation and optimising of contaminated sites remediation

7. General comparison of the three software tools

The RMK (REC) software and the environmental/economic evaluation software are based on the spreadsheet programme Excel ®. The Environmental balancing software works with the database programme Access ®.

Allowed input data are listed in Table 2. With several input data in all three software tools a wide variety of bioremediation options can be modelled.

Table 2: Type of input data from site investigation (no costs)

Type of input data	Environmental balancing	Environ./Econ. evaluation	REC
Mass of polluted soil (water)	X	X	X
Pollutant	X	X	X
Remediation technology	X	X	X
Time of remediation	X	X	X
Transport: mass and distance	X	X	X
Type of land use	X	X	X
Amount of materials (plastics, etc.)	X	X	
Pollutant concentration			X

The three software tools include predefined bioremediation processes. These predefined processes can consist of several other processes (ventilation, stirring, etc.).

The REC software and the environmental/economic evaluation include one soil bioremediation technology. Detailed descriptions of the applied technologies are not in the documentation of the two software tools and their documentation. If the software user knows the amount of electricity (REC), the amount of diesel (REC), the time of the usage of specific machines (environmental/economic evaluation) then the software user can model many bioremediation technologies. In case of the environmental/economic evaluation software the software user has to enter these data in several different sheets for the different phases of the remediation (establishment, operation, dismantling).

The environmental balancing software covers four different biotechnological technologies for the soil treatment (*in situ, ex situ*), two bioremediation technologies (*in situ* including reactive walls) and one technology of the biological cleansing of discharge air. It is also possible to model other bioremediation measures by using the basic technologies of civil engineering (well construction etc.). The key data of the technology (like amount of the soil) can be entered on one screen. The four predefined soil treatment technologies have been taken from one review publication (Franzius *et al.* 1996).

The environmental/economic evaluation software offers the possibility to enter data of the usage several dozens of (single) machines. In this respect, the environmental/economic evaluation software is the most detailed software and the REC software the least detailed one of the three analysed software tools. For many soil remediation-planning studies the detailed approach of the environmental/economic evaluation is not necessary because the uncertainty in the planning phase is often very high. In the REC software the software user should have some knowledge about the calculation of the electricity amount or diesel amount of the machines used. If the software user does not has this knowledge, the environmental balancing software can be more suitable.

8. Comparison of the life cycle assessment approaches

The REC method considers mainly diesel machine and electrical machine applications. These energy data are transformed directly into aggregated scores for air emissions. For example:

Input data	1 Gigagram contaminated soil in thermal soil treatment
Intermediate data	600 Gigajoule electricity
Output data	3 yearly Dutch inhabitant energy demand equivalents
	13 yearly Dutch inhabitant air emission equivalents

The assumptions and sources (literature) for the generic life cycle inventory data used are not documented.

Table 3: LCA output data of the three software products

Type of quantitative output data	Environmental balancing	Environ./Econ. evaluation	REC
Waste amount	X	X	X
Aggregated score for cumulated energy demand in Joule	X		
Up to 19 Disadvantage factors for single impact indicator results (inventory entries) for comparison of two remediation options	X		
Land use (only remediation measures)	X		
Water consumption (only remediation)	X		
Non-renewable resources indicator result	X		
Global warming indicator result	X	X	
Acidification indicator result	X	X	
Summer smog indicator result	X	X	
Toxicity indicator results (air, water, soil; human or ecosystem; persistent)	X	X	
Odour indicator results (near/remote emissions)	X		
Noise indicator results (different levels)	X		
Ozone depletion indicator result		X	
Nutrification indicator result		X	
Aggregated score for air emissions			X
Aggregated score for cumulated energy demand in yearly inhabitant equivalents			X

In the environmental balancing method, the result of the life cycle assessment approach is a table with up to 19 disadvantage factors. The environmental/economic evaluation gives tables with eleven traditional life cycle indicator results and 15 common inventory results (mainly resources and wastes). Eight of the eleven indicator results are associated with the environmental "costs" and three with the environmental "benefits". Most of these factors can be traced back to the indicator results (impact categories) and the single emissions (differentiated for the different processes like transportation by truck) or other inventory entries. The environmental balancing and the environmental/economic evaluation use mainly published generic life cycle inventory data. Some technologies (like the soil bioremediation) in the environmental/economic evaluation software are described with the emission data but not with the process data.

Table 3 shows output data concerning the three life cycle assessment approaches. Due to a missing of a clear separation between LCA data and non-LCA data in the three software tools, the definition of what can be interpreted as LCA data is taken by the distinctions made in the description about the Environmental balancing method (Volkwein *et al.* 1999).

9. Conclusions

All three software products include the evaluation of the environmental burdens of the remediation processes. All three software products help soil remediation planners to estimate the overall environmental performance of different remediation concepts. This is a major contribution for the inclusion of the sustainability concept in soil remediation planning.

All three software products are useful decision support tools complementing the soil remediation planning toolbox.

They require only data, which are usually available after a site investigation and a preliminary remediation concept making. All three are designed for soil remediation planners who have no (or little) experience in life cycle assessment.

The Environmental balancing software contains only calculation procedures for a life cycle assessment approach. The REC software includes additional calculation procedures for risk (reduction) assessments and cost assessments. Therefore, only the life cycle assessment approach can be compared between the three software products. The transparency of the generic data used (life cycle inventory) and assumptions and the transparency of the output data (life cycle impact assessment) are higher in the environmental/economic evaluation and environmental balancing than in the REC software.

The environmental/economic evaluation software allows detailed data inputs for specific machines. By avoiding these detailed information the environmental balancing helps the planner to focus on the important issues of the remediation options.

References

Beinat, E.; van Drunen, M. A.; Janssen, R.; Nijboer, M. H.; Kohlenbrander, J. G. M.; Okx, J. P.: The REC decision support system for comparing soil remediation options. A methodology based on risk reduction, environmental merit and costs. Civieltechnisch Centrum Uitvoering Research en Regelgeving (CUR), Nederlands Onderzoeksprogramma biotechnologische In situ Sanering (NOBIS), Büchnerweg 1, Postbus 420, 2800 AK Gouda, Netherlands 1997.

Bender, A.; Volkwein, S.; Battermann, G.; Kohler, W.: Umweltbilanz von Altlastensanierungsverfahren. TerraTech. Zeitschrift für Altlasten und Bodenschutz 1999, 8, 37 – 41.

Bender, A.; Volkwein, S.; Battermann, G.; Klöpffer, W.; Hurtig, H.-W.; Kohler, W.: Life cycle assessment for remedial action techniques: methodology and application. ConSoil '98, Sixth international FZK/TNO Conference on contaminated soil. Thomas Telford, London. 1998, 367 – 376.

Diamond, M. L.; Page, C. A.; Campbell, M.; McKenna, S.; Lall, R.: Life-cycle framework for assessment of site remediation options: Method and generic survey. Environ. Toxicol. Chem. 1999, 18, 788 – 800.

Doetsch, P.; Dreschmann, P.: Verfahrensdokumente zur mikrobiologischen Bodenbehandlung (On- und Off-site-Verfahren). In: Franzius, V.; Stegmann, R.; Wolf, K.: Handbuch der Altlastensanierung. R. v. Decker's Verlag, G. Schenck Heidelberg. 13. Lieferung 12/92, 1 – 41.

Franzius, V.; Wolf, K.; Brandt, E. (Editors): Handbuch der Altlastensanierung. R. v. Decker's Verlag G., 1996

Frischknecht, R.; Hofstetter, P., Knoepfel, I.; Dones, R.; Zollinger, E.: Ökoinventare für Energiesysteme. Bundesamt für Energiewirtschaft, Bern. 2. edition. ENET, Postfach 130, 3000 Bern 16, Switzerland 1995

ISO 14040:1997: Environmental Management. Life cycle assessment. Principles and framework. International Standards Organization, Geneva (www.iso.ch).

ISO 14041:1998: Environmental Management. Life cycle assessment. Goal and scope definition and inventory analysis. International Standards Organization, Geneva (www.iso.ch).

ISO 14042:2000: Environmental Management. Life cycle assessment. Life cycle impact assessment. International Standards Organization, Geneva (www.iso.ch).

ISO 14043:2000: Environmental Management. Life cycle assessment. Life cycle interpretation. International Standards Organization, Geneva (www.iso.ch).

Joziasse, J.; Bakker, T.; Eggels, P. G.: Decision support system for treatment of dredged sediments. ConSoil '98, Sixth international FZK/TNO Conference on contaminated soil. Thomas Telford, London. 1998, 1193 – 1194.

Landesanstalt für Umweltschutz Baden-Württemberg: Umweltbilanzierung von Altlastensanierungsverfahren. Version 1.0 Rev. 16. CDROM including description of methodology. 1999. Available from AHK Gesellschaft für Angewandte Kartographie mbH, Rehlingstraße 9, 79100 Freiburg, Germany, phone 07 61 – 70522 – 0 (costs below 50 Euro).

Lieber, M., Kobberger, G., Weingran, C.: Comprehensive impact assessment of site remediation (CIASR): Tool for planning, strategic controlling, valuations and decisions in the course of remediation. Proceedings of ConSoil 2000. 7th International FZK/TNO Conference on Contaminated Soil. Thomas Telford, London. 2000, 125 – 132.

LIFE: Environmental/Economic Evaluation and optimising of contaminated sites remediation. Report prepared for the Danish national Railway Agency and the Danish State Railways by HOH Water Technology A/S, NIRAS Consulting Engineers and Planners A/S, Revisoramvirket /Pannell Kerr Forster, ScanRail Consult. EU LIFE Project n. 96ENV/DK/0016 supported by Danish Environmental Protection Agency. CD-ROM available from ScanRail consult, Pilestraede 58, 3, 1112 Copenhagen, Denmark, phone +45 – 33 76 50 05 ext. 13623.

Nijboer, M. H., Schelwald-van der Kley, A. J. M.: Comparison of remediation technologies. Outcome of questionnaire and overview of developments. NICOLE. The Network for Industrially Contaminated Land in Europe. 1998. Tauw Milieu bv, Handelskade 11, Postbus 133, 7400 AC Deventer, The Netherlands.

van Drunen, M. A., Beinat, E., Nijboer, M. H., Haselhoff, A., 't Veld, M., Schütte, A. R.: De RMK-methodiek voor het beoordelen van bodemsaneringvarianten. Een methode gebaseerd op Risicoreductie, Milieuverdienste en Kosten – RMK fase 3. Internetversie. 12 April 2000 (free download: www.vu.nl/ivm/research/rmk).

Volkwein, S.: Governmental policy for soil remediation regarding overall environmental performance. Proceedings of ConSoil 2000. 7th International FZK/TNO Conference on Contaminated Soil. Thomas Telford, London. 2000a, 55 – 56.

Volkwein, S.: Comparison of software tools: "REC" and "Umweltbilanzierung von Altlastensanierungs-verfahren" Proceedings of ConSoil 2000. 7th International FZK/TNO Conference on Contaminated Soil. Thomas Telford, London. 2000b , 1397 – 1404.

Volkwein, S.: Life cycle assessment of soil remediations with a software tool. Proceedings of ConSoil 2000. 7th International FZK/TNO Conference on Contaminated Soil. Thomas Telford, London. 2000c, 1405 – 1406.

Volkwein, S., Hurtig, H.-W., Klöpffer, W.: Life Cycle Assessment of Contaminated Sites Remediation. Int. J. LCA 1999, 4, 263 – 274.

Volkwein, S. Klöpffer, W.: Bioreactor for soil treatment: Life cycle assessment. In: M. Hofman (Editor). Proceedings of the NINTH EUROPEAN CONGRESS ON BIOTECHNOLOGY 11 - 15 July 1999. Branche Belge de la Société de Chimie Industrielle - ISBN 805215-1-5 ECB9 CD ROM 1- 8. Available from caulca@yahoo.com.

Volkwein, S.; Bender, A.; Klöpffer, W.; Hurtig, H.-W.; Battermann, G.; Kohler, W.: Life cycle assessment method for remediation of contaminated sites. ConSoil '98, Sixth International FZK/TNO Conference on Contaminated Soil. Thomas Telford, London. 1998, 1069 – 1070.

Volkwein, S.; Klöpffer, W.; Hurtig, H.-W.: Umweltbilanz von Altlastensanierungsverfahren. Proceedings Ökobilanzen – Trends und Perspektiven . Working group Ökobilanzen of the Gesellschaft Deutscher Chemiker (Fachgruppe Ökologische Chemie). In the DECHEMA building in Frankfurt am Main, Germany. 26 June 1997.

PART 2
EX SITU CLEAN UP TECHNOLOGIES

THE DMT-BIODYN-PROCESS

A suspension reactor for the biological treatment of fine grained soil contaminated with polycyclic aromatic hydrocarbons

CHRISTOPH SINDER, JÜRGEN KLEIN AND FRANK PFEIFER
Deutsche Montan Technologie GmbH, Division GUC
Am Technologiepark 1, D-45307 Essen, Germany
E-mail: Sinder@dmt.de

Abstract

The objective of the research was to scale up, test and optimise the DMT-BIODYN process for the biotreatment of fine-grained, PAH contaminated soils. Therefore, a fluidised bed bioreactor (1 m³-scale) was designed and constructed. After prior investigations according to process engineering parameters and biodegradability, different PAH contaminated soils were successfully treated. The soil is fluidised by means of an upward liquid flow in a slurry reactor. The reactor configuration allows a high solid content (up to 50 % w/w), while a low energy input is sufficient to maintain the soil suspended. The bioavailability of the pollutants is optimal due to the separation of all soil particles. This results in short bioremediation periods (days to few weeks). PAH degradation rates of more than 95 % were achieved for some soils. The results obtained for PAH biodegradation in the pilot scale bioreactor were similar to those obtained in the laboratory system, which is working under optimal conditions. Based on the investigations carried out a further scale up of the process was possible. Today, the DMT-BIODYN process is operated in industrial scale.

1. Introduction

A significant portion of the known contaminated sites in the Federal Republic of Germany is accounted for by former coal by-product facilities in the industrial region of the Ruhr. The natural soil of this area in many cases has a high fine-grained content and is, at and around the abandoned sites, frequently polluted with organic substances, particularly polycyclic aromatic hydrocarbons (PAHs). Microbial degradation of PAHs has in the past been investigated in a large number of studies (Cerniglia and Heitkamp 1989; Weißenfels *et al.* 1990; Cerniglia 1992). These publications showed that many of

37

S.N. Agathos and W. Reineke (eds.).
Biotechnology for the Environment: Soil Remediation,37-49.
© 2002 *Kluwer Academic Publishers. Printed in the Netherlands.*

these substances are degraded by bacteria or fungi. The particular microorganisms utilize differing degradation strategies for this purpose, whereby the target of biological remediation of contaminated soil is generally the complete mineralisation of the PAHs (Kästner *et al.* 1993). A restricting parameter in this context has proven to be the low solubility of the PAHs in water, which decreases as molecular weight rises. The associated adsorption of the PAHs on to hydrophobic constituents of the soil can result in a considerable decrease in bioavailability (Karickhoff *et al.* 1979; Chiou *et al.* 1995).

A complicating factor in the case of fine-grained soils is the fact that the poor accessibility of the pollutants in these soils resulting from the severely limited physical access routes hinders the conventional biological remediation method in the form of heap-type processes. Nor are soil-washing processes suitable for treatment of such soils, since in this case cleaning is effected, among other ways, by shearing off the pollutants from the soil particles and it is not possible to apply sufficiently high shearing forces to small particles.

In view of the situation discussed above and the clear need for development work in this field, DMT has developed for biological remediation of organically contaminated fine-grained soils the DMT-BIODYN process (Nitschke 1994; Mann 1997), which fully conforms with modern economic and ecological requirements. The development targets were the complete contacting of all soil particles by the aqueous phase, the assurance of optimum habitat conditions for bacterial degradation of PAHs, a high bioreactor solids content and a low energy consumption.

2. Materials and methods

2.1. THE PRINCIPLE OF THE DMT-BIODYN PROCESS

In the DMT-BIODYN process, the soil is screened to a particle diameter of < 1 mm and the oversize particles are treated separately in another process. The central element in the process consists of treatment in a hydraulically fed bioreactor (bed-bed reactor). Figure 1 shows a simplified flow sheet for the DMT-BIODYN process.

The soil to be treated is hydraulically suspended in Reactor A1 by means of vertical approach flow and the individual soil particles are separated from one another. The expanding cross-section of the top of the reactor achieves size-classification and solids reduction of the recirculating suspension. In the aqueous phase, the organic pollutants are aerobically degraded by bacteria to, water and biomass. The necessary oxygen is fed into the soil suspension by means of external CO_2 gas feed in bubble column B1. Here, the suspension is saturated with oxygen, fed back to Reactor A1 by means of a recirculating pump, and used for fluidisation. In addition to the external aeration system, internal aeration directly in Reactor A1 can also be activated if oxygen consumption is high. The oxygen concentration on the feed and return sides of A1 is monitored and recorded. It is kept constant in the feed line by means of regulation of the volumetric inflow of aeration air. The pH value is monitored and can be adjusted by adding acid or alkali if necessary.

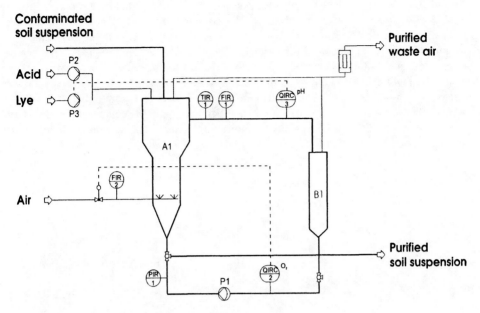

Figure 1: Simplified flow sheet for the DMT-BIODYN process

2.2. LABORATORY FACILITY

On the basis of preliminary investigations, a laboratory facility for the microbiological treatment of fine-grained soils was designed and used for the determination of fundamental process parameters. The bed-bed reactor features only one feed hopper and has a capacity of 10 l, of which 8.5 l are available as the effective working volume. Aeration of the soil suspension is accomplished externally in a bubble column with a fritted glass sparger (total capacity 2 l, working volume 1 to 1.5 l). Both reactors consist of glass.

The solids concentrations used within the scope of the tests for microbial degradation of PAHs were 50 % (w/w), both in the laboratory facility and in the pilot plant. All other experimental conditions (pH, nutrient salts, etc.) were identical to the tests in the Sapromat (see 2.4.).

2.3. PILOT PLANT

The pilot bed-bed reactor plant has an overall capacity of 1.4 m³, with an effective working capacity of 1.2 m³. The geometry of the new reactor was modified vis-à-vis the laboratory system in view of the later intended increase in scale to industrial proportions. In order to reduce the suspension reactor's height/diameter (h/d) ratio, the number of feed hoppers in the approach-flow zone was increased to four ("multiple feed hoppers"). The h/d ratio of each flow-cell, consisting of one feed hopper and the section of the reactor above it, remained constant. Two pneumatically operated diaphragm pumps were used for recirculation of the soil suspension. Hose-type aerators were used both in the bubble column (B1 in Figure1) and in the bed-bed reactor (A1 in Figure1).

An agitator for mechanical foaming-suppression was installed below the reactor cover in both the suspension reactor and the bubble column. The pilot plant is equipped with measuring systems for monitoring and regulation of oxygen content and pH value and for monitoring of temperature in the reactors. In addition, carbon dioxide and oxygen in the waste-gas are monitored on-line by means of an infrared spectrometer. Hydrocarbon content (qualitative analysis) in the waste air is measured using a standalone photoionization detector (PID) (Thermoinstruments GmbH, Dortmund, FRG) with a 10.0 eV lamp. Calibration is carried out using isobutylene. PAH content in the waste air is detected by HPLC after adsorption on amberlite-XAD16 (Fluka, Taufkirchen, FRG) and subsequent extraction of the adsorption material (see 2.5).

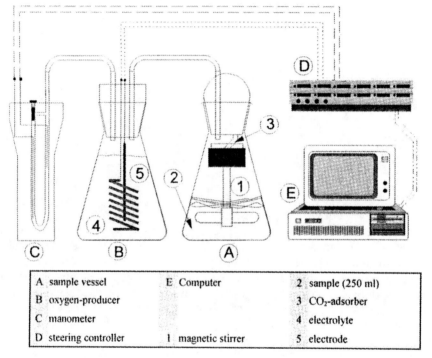

A sample vessel	E Computer	2 sample (250 ml)
B oxygen-producer		3 CO$_2$-adsorber
C manometer		4 electrolyte
D steering controller	1 magnetic stirrer	5 electrode

Figure 2: Sapromat system.

2.4. SAPROMAT SYSTEM

The Sapromat system (Sapromat D12E, Voith, Heidenheim, FRG) is a respirometer (Figure 2), which allows biological soil treatment as a suspension (stirred by a special mixer). In the course of biotreatment oxygen-consumption is measured continuously in the closed experimental system.

Soils were treated as a suspension with 10 % (w/v) of soil in mineral salt medium (MSM) (Walter *et al.*, 1991) and inoculated (5 % v/v) with a bacterial mixed culture able to degrade complex PAH-mixtures. This mixed culture was enriched on anthracene oil as the sole source of carbon and energy (Walter *et al.*, 1991). After incubation at

20°C the soil slurry was quantitatively removed and prepared as described below for chemical analyses.

2.5. CHEMICAL ANALYSIS

PAH analyses of soil suspensions were performed after centrifugation at 25,000 xg for 30 min. The soil was dried for 16 h at 40 °C and extracted with toluene by use of a Soxhlet extractor unit (Tecator, Höganäs, Sweden). Adsorption material (XAD) is extracted in the same way. The extracts were cleaned by passing through a silica gel column (Merck, Darmstadt, FRG), concentrated under a gentle stream of nitrogen and analysed by HPLC. The PAHs were separated on a LiChrosorb RP 18 column, 250 mm x 4.6 mm (Merck, Darmstadt, FRG), with a 40:60 acetonitrile/water - 100 % acetonitrile gradient, run over 30 min at a flow rate of 1 ml/min using a liquid chromatograph HP 1090 A with autosampler and photodiode array detector (Hewlett Packard, Waldbronn, FRG).

BTXE-aromatics and methylated naphthalenes were analysed by gaschromatography (Typ HP 5890, Fa. Hewlett-Packard, Düsseldorf, FRG; Chrompack 5000 x 0.5 mm CP-sil5 column, FID-detector) after the extraction of wet soil samples by acetone and reextraction of acetone by pentane.

2.6. PROCESS-ENGINEERING CHARACTERIZATION

The 3-s criterion was used with reference to Stiess (1992) as a criterion for suspension state. A bulk bed was considered to be completely suspended provided no particle remained in the approach-flow zone for more than 3 seconds.

The residence-time distribution was determined by means of the Peak Method (Murphy and Timpany 1967). A tracer (KCl) was fed into the bed-bed reactors' inlet as a point marker and the time-point of maximum tracer concentration in the outlet, measured on the basis of conductivity, was determined. The dispersion index/soil suspension index (Bodenstein number = Bo) as per Arceivela (1981), as a characteristic value for mixing and residence time behaviour, was then calculated from peak time using the following formula:

$$d = 1/Bo = 4.027 \cdot 10^{-2.09 \cdot (t_p/t_0)} \quad \text{for} \quad 0.3 < (t_p/t_0) < 0.8 \quad [1]$$

$$d = 1/Bo = 0.2 \cdot (t_p/t_0) \quad \text{for} \quad 0.03 < (t_p/t_0) < 0.3 \quad [2]$$

(t_p = peak time and t_0 = theoretical residence time)

3. Results

3.1. RESULTS OF PROCESS-ENGINEERING CHARACTERIZATION

Suspension tests with different fine-grained soils were performed in the laboratory and the pilot plant at various solids contents in order to determine the minimum approach-flow velocity v_L necessary for achievement of complete fluidisation in the bioreactor. The test results (Table 1) indicate a rise in the minimum necessary approach-flow velocity in the case of coarser soil material (characterized here by the diameter D_{50} in the soil particles) and at higher suspension solids-contents. The approach-flow velocities necessary for complete suspension in the bed-bed reactor of the pilot plant, referred to the reactor cross-section in the reactor's centre zone, are of an order of magnitude of 10^{-3} ms^{-1}. They are greater by a factor of 2 than the approach-flow velocities determined for the bed-bed reactor in the laboratory scale. It proved possible to suspend a bulk layer of approx. 4 meters (this equates to a solids concentration of around 50 % w/w) at an approach-flow velocity of the same order of magnitude as in the laboratory system. This made it possible to fulfil already in the pilot plant an important precondition for ultimate scale-up.

Table 1: Minimum approach-flow velocity (v_L in 10^{-3}ms^{-1}) for different soils used in suspension tests with the laboratory and the pilot plant (- = not determined)

Solid content [% w/w]	10	30	50	10	30	50
Approach-flow zone	Single feed hopper			Multiple feed hopper		
Laboratory plant						
Soil 6 (d_{50}=23 μm)	-	-	0.39	-	-	-
Soil 2 (d_{50}=30 μm)	-	-	0.57	-	-	0.58
Soil 5 (d_{50}=32 μm)	0.35	0.43	0.60	0.42	0.49	0.62
Soil 1 (d_{50}=35 μm)	0.41	-	0.64	-	-	0.68
Soil 3 (d_{50}=45 μm)	-	-	0.88	-	-	-
Soil 6 (d_{50}=65 μm)	0.82	-	1.28	-	-	-
Pilot plant						
Soil 2 (d_{50}=30 μm)	-	-	-	-	-	1.15
Soil 5 (d_{50}=32 μm)	-	-	-	0.91	1.10	1.20
Soil 1 (d_{50}=35 μm)	-	-	-	-	-	1.24

No upper limit of the fluidised bed was observable in suspension of the soils. The broad range of particle-size distribution resulted in size-classification of the soil particles across reactor height. Coarse particles (> 250 μm) accumulated, as could also be clearly visually perceived, in the approach-flow zone. The suspension in the overflow section to the bubble column, on the other hand, mainly contained fine particles. Complete size-classification of the soil is prevented by recirculating suspension (fine particles) and turbulence in the reactor, particularly above the approach-flow zone.

The achievement of a good habitat for the microorganisms necessitates an adequate oxygen and nutrient supply by means of uniform distribution across reactor volume. In addition, thorough mixing is also necessary for rapid and efficient pH adjustment. The soil suspension index (Bodenstein number) at various solids concentrations and for various approach-flow geometries was determined, initially in the laboratory facility and then in the pilot plant, as an index of mixing efficiency in the bed-bed reactor. The hydraulically fed bed-bed reactor behaves in reaction near to the ideal mixing conditions in the theoretical continuous stirred tank reactor (CSTR). Mixing efficiency in the bed-bed reactor increases with rising solids concentration and/or rising approach-flow velocity (see Table 2). In the laboratory system, the type of feed hopper (single/multiple) does not cause any change in mixing performance.

Table 2: Comparison of the Bodenstein number (Bo) for soil 1 used in suspension tests with the laboratory and the pilot plant velocity (v_L = minimum approach-flow velocity)

Approach-flow zone	Single feed hopper		Multiple feed hopper	
	v_L	Bo	v_L	Bo
medium	$[10^{-3}$ ms$^{-1}]$	[-]	$[10^{-3}$ ms$^{-1}]$	[-]
Laboratory plant				
water	0.61	5.01	0.61	4.61
water	3.22	4.13	3.24	4.22
10 % w/w soil	0.61	4.12	0.61	4.01
30 % w/w soil	0.61	2.51	0.61	2.73
50 % w/w soil	0.61	0.71	0.61	0.64
50 % w/w soil	3.23	0.17	3.24	0.25
Pilot plant				
water			1.23	3.02
water			4.20	2.71
10 % w/w soil			0.91	2.51
30 % w/w soil			1.12	1.64
50 % w/w soil			1.24	0.82
50 % w/w soil			4.21	0.33

An explanation for the increase in mixing efficiency as solids concentrations rise can be found in the occurrence of greater turbulence in the bed-bed reactor as a result of an increased relative velocity between the particles and the fluid. This occurs at the same minimum approach-flow velocity and rising solids concentration, since the solids reduce the clear cross-sectional area of the reactor.

The soil suspension indices (Bodenstein number) for the pilot plant were of the same order of magnitude as had been determined for the laboratory system. At scale-up, it proved possible to maintain in the pilot plant the good mixing performance achieved in the laboratory system. This was intended to ensure in the pilot plant, too, adequate distribution in the bed-bed reactor of oxygen, nutrients and the acid/alkali added for pH adjustment.

3.2. RESULTS OF PAH-DEGRADATION EXPERIMENTS

The process-engineering characterization of the pilot plant shows that the preconditions for attainment of an optimum habitat for microbial degradation of PAHs were fulfilled in scale-up of the bioreactor, with assurance of complete suspension, thorough mixing and an adequate input of oxygen. This was then to be verified by means degradation tests on fine-grained soils.

For this purpose, Soils 1 and 2 were biologically treated both in the respirometer, the laboratory system and in the pilot plant. Soil 1 is essentially contaminated, at 223 mg PAH/kg soil, with 2-4 ring PAHs, whereas Soil 2, at a total of 1076 mg PAH/kg soil, also exhibits a significant 5/6 ring-PAH content (Table 3).

Table 3: Physico-chemical characteristics of the soils used for biodegradation experiments

Parameter			Soil	
			1	2
2-ring-PAH		[mg/kg]	101	21
3-ring-PAH		[mg/kg]	109	473
4-ring-PAH		[mg/kg]	12	411
5-ring-PAH		[mg/kg]	1	132
6-ring-PAH		[mg/kg]	< 0.5	38
Σ PAH$_{EPA}$		[mg/kg]	223	1 076
Particle size distribution				
gravel	> 2000 µm	[%]	< 0.1	< 0.1
coarse sand	630-2000 µm	[%]	0.5	0.6
middle sand	200-630 µm	[%]	0.5	4.2
fine sand	63-200 µm	[%]	28.7	18.5
coarse silt	20-63 µm	[%]	45.1	51.4
middle silt	6-20 µm	[%]	16.2	13.3
fine silt	2-6 µm	[%]	4.0	5.9
clay	< 2 µm	[%]	5.0	6.1

Both in the laboratory and pilot-scale systems, and also under the optimum conditions in the respirometer, the PAHs in Soil 1 are rapidly degraded without a noticeable "lag" phase (Figure 3). The rate of PAH degradation drops considerably after around 24 h. The rate of oxygen consumption correlates with the decrease in the PAH-degradation rate. A PAH concentration below the targeted value of 20 mg/kg$_{TS}$ is achieved after a treatment period of approximately 6 days. The decrease in PAH concentration with time is virtually identical at all three reactor scales and also correlates to an increase in the bacteria count in the soil suspensions. Fungi were not detected in the test materials. Not only the plots of PAH concentration and oxygen consumption but also all the other variables characteristic of bacterial degradation of PAH were equivalent in the laboratory and pilot plant systems to those of the degradation tests performed under ideal conditions for PAH degradation in the respirometer (Table 4).

The lower oxygen consumption in the experimental systems compared to the tests under optimum conditions in the respirometer can be explained by losses of readily volatile PAHs caused by stripping by the injected compressed air. 26 mg PAHs and 18 mg PAHs respectively were thus found on the XAD filters installed in the waste-air line.

Results virtually identical to those from the tests performed under optimum conditions in the respirometer were also achieved in PAH degradation tests in the laboratory- and pilot plant-scale tests systems on the more heavily contaminated Soil 2. In this case, PAH degradation is characterized by two phases (Figure 4).

The PAH concentration is initially reduced quickly, with a high oxygen consumption, as a result of degradation of the readily available 2- and 3-ring aromatics.

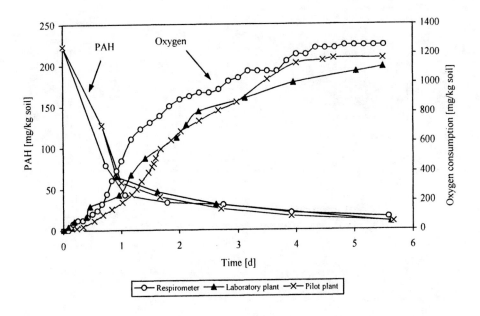

Figure 3: Course of PAH degradation and oxygen consumption for degradation experiments with soil 1 using different reactor systems

After three days, and the largely completed degradation of the easily degraded PAHs, the rate of PAH degradation decreases considerably. The decline in PAH concentrations is practically identical at all three reactor scales.

In the treatment of Soil 2 in the laboratory- and pilot plant-scale systems, all other variables characteristic of bacterial degradation of PAHs, as well as the plots for PAH concentrations and oxygen consumption, were also identical to those from the degradation tests performed in the respirometer (Table 5). The results of this degradation test also illustrate the success of the scaling-up of the biological suspension reactor process.

Table 4: Results of PAH-degradation experiments with Soil 1 in different reactor systems

parameter		Respirometer	Laboratory plant	Pilot plant
period of treatment	[h]	136	136	136
PAH-concentration				
start	[mg/kg$_{TS}$]	223	223	223
end	[mg/kg$_{TS}$]	16	10	11
abiotic losses [1]	[mg/kg$_{TS}$]	< 5	26	18
degradation rate	[%]	93	96	95
max. O$_2$ consumption rate	[mg/kg$_{TS}$·	52	49	50
point of maximum	[h]	25	28	37
total O$_2$ consumption	[mg/kg$_{TS}$]	1,258	1,107	1,168

[1] determined in the sterile control of the respirometer or the XAD-filters of the bioreactors

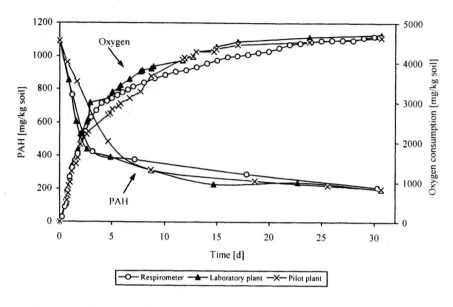

Figure 4: Course of PAH-degradation and oxygen consumption for degradation experiments with Soil 2 using different reactor systems

Table 5: Results of PAH-degradation experiments with Soil 2 in different reactor systems

parameter		Respirometer	Laboratory plant	Pilot plant
period of treatment	[h]	727	735	735
PAH-concentration				
start	[mg/kg$_{TS}$]	1 076	1 076	1 076
end	[mg/kg$_{TS}$]	209	197	201
abiotic losses [1]	[mg/kg$_{TS}$]	< 5	17	11
degradation rate	[%]	80	82	81
max. O_2-consumptionrate	[mg/kg$_{TS}$·h]	91	82	86
point of maximum	[h]	42	46	53
total O_2-consumption	[mg/kg$_{TS}$]	4,664	4,713	4,631

[1] Determined in the sterile control of the respirometer or the XAD-filters of the bioreactors

4. Discussion

Since, under suitable conditions, microorganisms are capable in principle of mineralising PAHs and a large range of other organic pollutants, biological treatment of contaminated soil is ecologically desirable (Cerniglia 1992). Biological treatment of fine-grained soils using conventional heap-methods has up to now failed due to the lack of biological accessibility of the pollutant substances, which results from the absence of physical access conditions for oxygen and nutrients (Kästner *et al.* 1993). The development objective of the DMT-BIODYN process was therefore that of creating and optimising by means of technical provisions the preconditions for microbial degradation of PAHs and other pollutants (Nitschke 1994; Mann 1997).

Energy is required for the fluidisation and maintenance of a soil suspension in a bioreactor. Despite the fact that treatment times are short compared to conventional biological remediation processes, this must nonetheless be taken into account in the selection of a suitable bioreactor. In addition, it is necessary to include investment costs for bioreactors and for the necessary control and instrumentation systems in planning, with the result that high solids concentrations in the reactor and short treatment times must be targeted. It was possible with the DMT-BIODYN process, thanks to the vertical hydraulic feed arrangement, to achieve and maintain complete suspension of all soil particles with solids contents of up to 50 % (w/w) with a low energy input at both laboratory and pilot plant scale. The achievable solids content is thus significantly higher than in other suspension reactors, such as Airlift and agitator reactors (approx. 25 to 30 %). The minimum theoretical energy input for suspension of this solids concentration is around 50 W per 1000 kg soil corresponding to an average energy requirement per day of treatment of 7 kWh per 1000 kg soil. In addition, it was also possible to achieve complete suspension of the soil at the maximum bulk bed depth used of approximate 4 m.

Oxygen input proved in the degradation tests to be sufficiently high. Since the amount of oxygen fed into the bioreactor depends on the minimum volumetric flow of suspension to be maintained, additional internal aeration is provided for treatment phases with exceptionally high oxygen consumptions. This makes it possible, unlike the situation in other suspension reactors, such as rotary reactors, for example, to ensure the microorganisms' supply of oxygen at all times by means of the O_2 input and the homogeneous mixing achieved. The rate of biological degradation of the pollutants is also dependent on the distribution of the nutrients, the O_2 and the substrate throughout the bioreactor. The tests performed in the laboratory and pilot plant systems indicated thorough mixing of the soil suspension in the reactor. In reaction terms, the bed-bed reactor behaves like a continuous stirred tank reactor (CSTR), with the result that effective pH regulation in the soil suspension is also possible. Compared to other reactors, the reactor type developed and the type of suspension selected provide the optimum process-engineering basis for efficient biological remediation of fine-grained soils.

This is confirmed by the degradation tests carried out with a range of PAH-contaminated soils in the pilot plant system, in which PAH content decreased significantly within a few days. Despite the significantly higher solids concentration, the necessary treatment times and the final PAH concentrations achieved are comparable to those obtained under ideal conditions in biological degradation tests in the Sapromat. Under these conditions, bacterial aerobic degradation of PAHs in soil suspension is not dependent on the scale of treatment.

The respirometer and the laboratory system are suitable for preliminary investigation of the biological remediability of fine-grained soils contaminated with PAHs. A reliable forecast of treatment time, achievable residual PAH concentrations and oxygen consumption rate can be obtained from the respirometer. In the laboratory system, the suspension behaviour of the soil can be studied and, if necessary, a characteristic particle diameter determined. The results of the PAH-degradation tests on Soil 2 simultaneously illustrated the fundamental problems of biological degradation of PAHs in soil. Despite complete suspension of all soil particles and the associated optimisation of the mass transfer processes, a significant residual PAH content remains in the soil. As could be demonstrated in other investigations not discussed here in more detail, adsorption of the PAHs on to the organic matrix of this soil results in a reduction of the biological availability of the PAHs. A conceptual solution for the cleaning of soils contaminated with non-biologically available PAHs is provided by the combination of treatment in a bioreactor with soil washing. It was possible in initial laboratory tests to achieve a significantly better degree of PAH removal (around 10 mg/kg soil) from Soil 2 than had been attained with either of the individual processes (DMT-BIODYN reactor and soil washing) (Sinder *et al.* 1994; Mann *et al.* 1996).

The investigations performed here prove the existence of the scientific preconditions for the next step, of scaling-up of the biological suspension reactor process from the pilot plant to the commercial scale. Geometrical optimisation of the approach-flow zone of the suspension reactor makes it possible to achieve a further enlargement of capacity, essentially by means of an increase in the diameter of the reactor, while retaining the geometrical similarity of the flow-cells.

Successful scale-up of the DMT-BIODYN reactor to industrial proportions has now been demonstrated on the basis of cooperation with a Swedish associate company (EkoTec AB, Skelleftehamn). Following the design and construction of the first demonstration plant (capacity 55 m^3), the first trial runs have been successfully completed.

Acknowledgements

The work was supported by the Bundesministerium für Bildung, Wissenschaft, Forschung und Technologie (BMBF 1480891).

References

Arceivala, S.J. (1981) Wastewater treatment and disposal, Marcel Dekker Inc., New York

Cerniglia, C.E. (1992) Biodegradation of polycyclic aromatic hydrocarbons, Biodegradation 3, 351-368.

Cerniglia, C.E. and Heitkamp, M.A. (1989) Microbial degradation of PAH in the aquatic environment, in U. Varasani (ed.) Metabolism of polycyclic aromatic hydrocarbons in the aquatic environment, CRC Press Inc., Boca Raton, pp 41-68.

Chiou, C.T., Kile, D.E., Zhou, H., Li, H. and Xu, O. (1995) Partition of non-polar organic pollutants from water to soil and sediment organic matters, Environ Sci Technol 29, 1401-1406.

Karickhoff, S.W., Brown, D.S. and Scott, T.A. (1979) Sorption of hydrophobic pollutants on natural sediments, Water Research 13, 241-248.

Kästner M. Mahro B and Wienberg R (1993) Biologischer Schadstoffabbau in kontaminierten Böden, in R. Stegmann (ed.), Hamburger Berichte Band 5, Economia Verlag, Bonn

Mann, V.G., Sinder, C., Pfeifer, F. and Klein J. (1996) Scale up und Optimierung eines Bioreaktorverfahrens zur Behandlung PAK-kontaminierter, feinkörniger Böden, in R. Stegmann (ed.), Hamburger Berichte 10 „Neue Techniken zur Bodenreinigung", Economica Verlag, Bonn, pp 435-446

Mann, V.G. (1997) Optimierung und Scale up eines Suspensionsreaktorverfahrens zur biologischen Reinigung feinkörniger, kontaminierter Böden. D.C. Hempel (ed.), ibvt Schriftenreihe der TU Braunschweig, FIT-Verlag, Paderborn, FRG.

Murphy, K.L. and Timpany, P.L. (1967) Design and analysis of mixing for an aeration tank, J. of the Sanitary Division, Proceedings of the American Society of Civil Engineers, 1-15.

Nitschke, V. (1994) Entwicklung eines Verfahrens zur mikrobiologischen Reinigung feinkörniger, mit polyzyklischen aromatischen Kohlenwasserstoffe belasteter Böden, PhD Thesis, Universität-Gesamthochschule Paderborn, Fachbereich Chemie und Chemietechnik, FRG.

Sinder, C., Fuisting J. and Klein J. (1994) Feinkörnige Böden sanieren: Bodenwäsche kombiniert mit biologischer Behandlung, Umwelt 5, 232 – 234.

Stieß, M. (1992) Mechanische Verfahrenstechnik, Springer Verlag, Berlin.

Walter, U., Beyer, M., Klein, J. and Rehm, H.J. (1991) Degradation of pyrene by Rhodococcus sp. UW1. Appl Microbiol Biotechnol 34, 671-676.

Weißenfels, W.D., Beyer, M. and Klein, J. (1990) Degradation of phenanthrene, fluorene and fluoranthene by pure bacterial cultures, Appl Microbiol Biotechnol 32, 479-484.

THE SLURRY DECONTAMINATION PROCESS

Bioprocessing of contaminated solid waste streams

R.H. KLEIJNTJENS, L. KERKHOF, A.J. SCHUTTER, K.CH.A.M. LUYBEN, J.F. DE KREUK AND J. JANSE
RHK: SGS-Ecocare, Planetenlaan 2, 3301 CE Dordrecht, Netherlands, Email: Rene_kleijntjens@sgsgroup.com; LK: Kerkhof and Zn, Rotterdamseweg 452, 2629 HJ, Delft, Netherlands; AJS: P.I.M. (Partners In Milieutechniek, Mercuriusweg 4, 2516 AW Den Haag, Netherlands; KChAML: Faculty of Applied Science, Technical University of Delft, Julianalaan 134, 2628 BL Delft, the Netherlands; JFdK & JJ: Biosoil, Nijverheidsweg 27, 3341 LJ Hendrik Ido Ambacht, Netherlands

Summary

Within the broad spectrum of technologies, focusing on the treatment and recycling of contaminated soils, sediments and sludge four types of *ex situ* bioprocessing can be identified:

- landfarming

- composting (biopiles)

- solid state fermentation (rotating reactors)

- slurry processing

To the latter category belongs the Slurry Decontamination Process (SDP), the SDP is an *ex situ* continuous plug flow system based on tapered airlift slurry bioreactors. The process consists out of 4 major unit operations in which separation technology and biotechnology are integrated. The system has been studied at scales of 400 l, 800 l and 4 m^3.

The integral process was operated at pilot scale (3 m^3 reactor volume) to test various solids waste streams. It was operated semi-continuously over a period of two and a half years. In the SDP efficient sand removal was combined with steady state microbial

S.N. Agathos and W. Reineke (eds.).
Biotechnology for the Environment: Soil Remediation. 51-70.
© 2002 Kluwer Academic Publishers. Printed in the Netherlands.

breakdown of organic pollutants such as mineral oils, PAH, BTEX and to a lesser extend PCB's. Recycling targets were reached. Major process parameters are the type and level of the contaminants, the power input, the solids hold-up, the power, capital and labor costs. The solids residence time is the key design parameter. Model calculations were made for a 1200 cubic meter scaled-up SDP-installation. For residences times ranging from 2 days up to 12 days, the costs were estimated. For the short residence times the cost level was within the market range (roughly be in between 20 and 60 euro/tonne), longer residence times were not feasible

In terms of scale up economics, the solids residence time in the reactors of the SDP should not be longer then strictly necessary (few days). Therefore reactor treatment (together with sand removal and flotation if required) is to be combined with low cost, extensive, after treatment such as ripening fields, landfarming, phytoremediation or biopiles (the so-called "constructed natural systems"). Due to the integration of intensive reactor treatment and extensive after treatment, not only recycling can take place within economic restrictions, but also flexibility is introduced to treat cocktails and solids waste streams having various compositions.

1. Introduction

1.1. RECYCLING OF CONTAMINATED SOLID WASTE

Waste recycling plays a key role in the development of a sustainable economy (Suzuki 1992). The classical approach, remediation without the production of recycled materials, does not contribute to durable material flows. Moreover, the production of

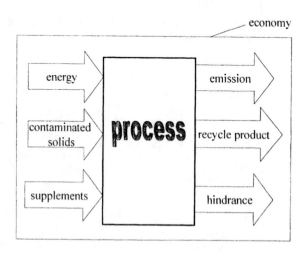

Figure 1: Elements of the environmental process balance

reusable materials is a necessity to make waste treatment an attractive economic solution. For sustainable solutions the overall environmental efficiency should be positive and fit within the local economic and legal framework (RIZA 1997). The benefits of the treatment (decontamination and recycling) thus have to be compared to the disadvantages (power use, emissions, hindrance). Since recycling cannot take place regardless the effort and costs needed, the environmental benefits finally are to be judges by their economic merits (Figure 1).

Solids waste streams (contaminated soils, sediments and sludge) can be recycled. The solids have to be recycled into usable products while the contaminants are removed or destroyed. In the Netherlands legal targets for recycled materials were set in the Dutch Building Material Decree (Staatsuitgeverij 1995). Depending on the content and the leaching of components, two different ways for using recycled materials in works are defined:

- Category 1 products, needing no further isolation

- Category 2 products, needing further isolation and monitoring

In the Dutch practice solids treatment is nowadays mainly aiming at the production of Category 1 recycle products. Although each country still has its own standards, and standardization is far away (Northcliff *et al.* 1998), it is beyond discussion that both recycling targets and environmental efficiency analyses are needed to support the development of sustainable technologies.

1.2. CHARACTERISTICS OF CONTAMINATED SOIL, SEDIMENTS AND SLUDGES

A complex solid matrix contaminated with one or more pollutants is the core of polluted soils, sediments and sludge (industrial or municipal). In soils the solid matrix is frequently dominated by sand, while the water content may be less then 25 %; levels of debris can be found depending on the history of the site but rarely exceed the 10 %. River, harbour and canal sediments contain a majority of water (frequently above 60/70 %) while the fine fraction (below 63 micrometer) may dominates the solids. Industrial and municipal sludge mostly are very humid (more then 95 % water) and have a large content of organics (above 60 %) (INES 1997). In industrial sludge the contaminant mostly originating directly from corrosion or wear (e.g. spent catalyst) of the installation on the site.

Disregarding the heterogeneous nature of the waste, the contaminant behaviour largely is determined by the fines (Apitz *et al.* 1994), This is due to the fact that submicron particles such as humic-clay structures and clay agglomerates have extreme high adsorption capacity (Brady 1984). The solid waste therefore basically contains a contaminated fine fraction, a less contaminated sand/ gravel fraction, cleaner debris and a contaminated water phase.

Corresponding to this broad spectrum of waste streams with their contaminants, there is a variety of treatment options.

2. Classification of treatment technologies

Basically solids waste treatment can be classified in three groups: *in situ* remediation, constructed natural systems (simple technology) and *ex situ* technology (Table 1).

Table 1: Classification of solids treatments

In situ remediation	Constructed "natural" technology / Simple processes systems	*Ex situ* remediation
No excavation No power input	Limited power input Low complexity Use of "natural processes"	Excavation Power input Process operations
Soil	Soil/Sediments	Soil/sediment/sludge
Pump and treat Bioremediation Soil venting Soil washing	Landfarming (soil) Ripening (sediment) Sand removal in sedimentation basins Aerobic lagoon treatment Anaerobic landfill treatment	Separation technology Thermal treatment Chemical/physical treatment Immobilisation Bioreactors

2.1. *IN SITU* REMEDIATION

In-site bioremediation basically is used within the urban context of contaminated soils when site conditions does not allow for excavation. A variety of *in situ* techniques have been developed in the US and in Europe to decontaminate soil without removal of the buildings above (Staps 1990). *In situ* treatment includes electro-chemical approaches (electrodes in the soil), soil sparging or venting, soil washing with water, or bioremediation stimulating the local biosystems. Another option is to "pump and treat" the groundwater present on the site; this means that the contamination is treated by means of treating the groundwater. *In situ* technology has been widely applied, an internet search carried out in February 1999 resulted in 941 hits for the term "*in situ* remediation".

2.2. CONSTRUCTED "NATURAL" SYSTEMS/SIMPLE TECHNOLOGIES

The approach notified as "constructed natural systems" or "simple technology" combines features of natural occurring processes and technological principles. An examples is landfarming used for the treatment of contaminated soils. The contaminated soil is gathered in heaps or layers in a controlled zone in which microbial breakdown takes place; forced aeration, the recycling of percolation water, addition of nutrients are ways to enhance the degradation rate (Riser-Roberts 1997). Although theoretically landfarming might be considered as a "bioreactor", in practise it is an open system without clear system boundaries. Landfarming therefore typically has the features of a "constructed natural system".

Closely related to landfarming is the "ripening" of contaminated sediments. Layers of contaminated sediments are spread out in a controlled zone; under the influence of the processes such as evaporation and natural dewatering an improvement of texture and composition of the ripened material is achieved.

Phytoremediation is using specific vegetation in relationship to landfarming and sediment ripening, two pathways can be followed:

- the plants or trees are used to accumulate metal contaminants such as copper, zinc, cadmium and nickel. Species which have been used in this respect are flax, maize and miscanthus (Kerr *et al.* 1998)

- the root structure (rhizosphere) of the growing vegetation is used as a matrix, which will stimulate the microbial remediation of organic, contaminates.

After the treatment the crop can be harvested and used as a sustainable biomass fuel (Harmsen *et al.* 1999); after incineration the contaminants are gathered in the ash or in the off gas filtration unit.

A simple separation technology making use of the "natural" gravity is the sedimentation basin for the treatment of sediment slurries. A sedimentation basin is a natural clarifier that provides conditions to allow suspended particles to settle out of a slurry. In this way sand is captured at the bottom of the basin while the overflow of contaminated fine fraction may be further treated (Cullinane *et al.* 1990; Van Leeuwen *et al.* 1997).

A more advanced "natural system" is the treatment of contaminated sediments in waste lagoons. Using air spargers at the lagoon bottom a particle suspension can be sustained in which microorganisms can aerobically degrade the (organic) contaminants (EPA 1994). Closely related to lagoon treatment are the processes focusing on the anaerobic microbial breakdown in sediment disposal sites.

2.3. *EX SITU* PROCESSING

Ex situ processes are characterized by the presence of:

- the solid waste as feedstock is pre-treated

- power input

- use of unit operations

A class of frequently used *ex situ* remediation options, especially for complex contaminated waste cocktails, is thermal treatment. As a follow up of soil treatment, thermal techniques also are being developed to treat (wet) sediments and sludge. Immobilisation of the solids into end products such as recycled gravel, bricks, of larger structures such as basalt have been established (Schotel *et al.* 1997).

A second class of successfully applied *ex situ* processes is based on particle separation techniques mostly originating from the field of mineral ore processing.

Common operations are sieves, flotation cells, hydrocyclones, Humphrey-spirals, jigs, fluid bed systems and up flow columns (Cullinane *et al.* 1990; Perry and Green 1984). These processes typically result in the removal debris and the production of reusable sand fractions while the contaminated fines are further treated or stored.

Chemical treatments, such as solvent extraction methods (e.g. supercritical CO_2 or acetone extraction) have been investigated as *ex situ* treatment technology. Also the addition of chemicals to solidify the solids is in practice being carried out (Hinsenveld 1995). Although applied in the US and Canada chemical solidification is no common use in Europe.

In case the contaminants are organic (such as mineral oil, PAHs, solvents, BTEX, PCBs) bioreactors can be used. In bioreactors populations of soil-organisms degrade the contaminants to yield harmless products (Schegel 1986).

3. Bioreactors

Defining a bioreactor as a vessel or a closed system in which under controlled conditions microbial breakdown occurs, the basic kinetics can be denoted by means of the black box model as depicted in Figure 2. Under steady state condition (no accumulation in the system), the input of pollutant, oxygen, nitrogen and soil compounds (e.g. humic substances) results into the output of carbon dioxide, water, nitrate and protons (Kleijntjens 1991):

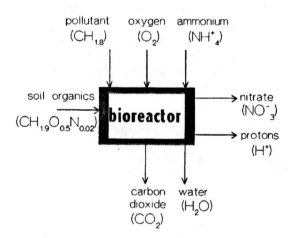

Figure 2: Basics of the bioreactor conversion

Although the mass balances for each of the elements involved (C, H, O and N) in theory should fit for the complete 100% (elements can not be destroyed), in practical multi-phase systems proper mass balance measurements are difficult. Besides interference with the surrounding for the gaseous components (e.g. oxygen), the dissolved components have to be followed in a complex matrix (Lotter *et al.* 1990). In practice the microbial breakdown therefore mostly is followed by the contaminant concentration only.

Having defined the major microbial features of the bioreactor, two major categories can be distinguished:

- bioreactors with a restricted solids hold-up: slurry reactors (typical solids hold-up below 40 wt%)

- bioreactors with restricted humidity: solid-state fermentation, (solids hold-up typical above 60 wt %)

3.1. SLURRY BIOREACTORS

Characteristic for all types of slurry bioreactors (Figure 3) is the need of power input to sustain a three phase system in which the solid particles are suspended; the gravity forces acting on the solids have to be compensated by the drag forces executed by the liquid motion (Hinze 1959). In a proper designed slurry system the power input is used to maintain three phenomena:

- suspension

- aeration

- mixing

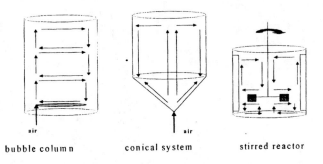

bubble column   conical system   stirred reactor

Figure 3: Common configurations of slurry bioreactors

For each reactor configuration, the appropriate balance of these three parameters depend on the reactor scale, the particle size distribution, slurry density, slurry viscosity, oxygen demand of the biomass and the solids hold-up (Kleijntjens 1991, Cheremisinoff 1986).

3.2. SOLID STATE BIOREACTORS

For solid-state fermentation there is no need to maintain a solids/liquid suspension; a compact moist solid phase determines the system. In a solid-state fermentor, process conditions are maintained by controlling the temperature, humidity and aeration. Both the fixed bed reactor as well as the rotating drum bioreactor are suited for solid-state fermentation (Figure 4). In the fixed bed reactor the contaminated solids are permanently installed upon a drained bottom as a stationary phase. Forced aeration and the supply of water mostly are applied as continuous phase (Riser-Roberts 1997). A fixed bed reactor may be batch operated as a closed biopile system.

Continuous solid state processing is possible in the rotating system, here the solid phase (as a compact moist material) is "screwed and pushed" through the reactor. In line with slurry processing power is required to maintain the transport of the solids through the system (Kiehne *et al.* 1995).

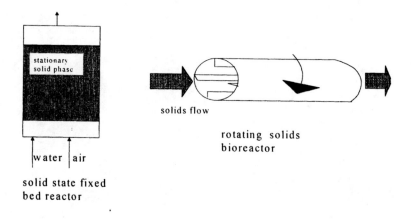

stationary
solid phase

solids flow

rotating solids
bioreactor

water   air

solid state fixed
bed reactor

Figure 4: Bioreactors for solid state processing (fixed bed and rotating drum reactor)

In line with its traditional role as a soil fertilizer, compost has been used as an additive in solid-state treatment. Basically the compost addition is used to stimulate the microbial breakdown. In experiments soil contaminated with hydrocarbons has been mixed with compost in various ratios (soil/compost ratios 2:1, 3:1 and 4:1). In 3-liter test batch reactors the hydrocarbon degradation was more than 90 % after a period of 44 days. Compared to the breakdown result without compost addition, the soil/compost systems showed a much faster degradation rate and a lower end concentration (Lotter *et al.* 1990).

At larger scale fixed bed experiments, using 10 m^3 biopiles, were carried out to investigate the degradation of chlorophenol in contaminated soil. Chalk, commercial fertilizer (NPK) and bark chips (as bulk aeration agent) were added as supplements. After two months 80 % of the contaminant was removed (Laine and Jorgensen 1995).

4. Configuration of *ex situ* bioprocesses

Depending on the type of bioreactor pre- and post-treatment operations are needed. Solid-state bioreactor processing in general does not need intensive pre-treatment of the solids since the texture of the solid feedstock is not significantly changed. In contrast, slurry bioprocessing requires an extensive pre-treatment to remove large parts from the solids before the reactor is being fed. The bioreactor mostly is integrated with washing/separation operations (Robra *et al.* 1998).

4.1. *EX SITU* SLURRY BIOPROCESS

A typical set-up of an integrated *ex situ* (slurry) bioprocess is depicted in Figure 5. First, the feedstock is screened using a wet vibrating screen, to remove the debris (sizes above 2-6 mm). Second, sand fractions are being removed by one or more separation techniques, a typical separation diameter (the so called "cutpoint") for these separation steps is 63 microns (or 50 microns depending on the standard chosen). In the cyclone shown in Figure 5 the slurry flow is split into a sand fraction (particle size above 63 microns) at the bottom and a fine fraction at the top (below 63 micrometer).

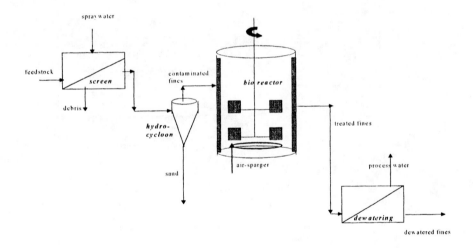

Figure 5: Typical set-up of an ex situ (slurry) bioprocess using a batch operated aerated stirred tank reactor (typical solids hold-up is 20 wt%)

The top flow of the cyclone, containing the contaminated fines, is being fed the bioreactor (depicted is a stirred tank but any of the three types from Figure 3 might be chosen). The final dewatering operation results in an end product containing the fines and process water.

Two types of *ex situ* bioprocesses can be identified:

- batch operation; no fresh material is introduced to the bioreactor during processing, the composition of the content changes continuously;

- (semi-) continuous operation (plug flow); fresh material is introduced and treated material removed during processing, the composition in the reactor remains unchanged with time (Levenspiel 1972);

Although continuous processes offer many advantages in terms of capacity, inactive periods and costs, in the treatment practice most operations are still batchwise.

4.2. SLURRY DECONTAMINATION PROCESS

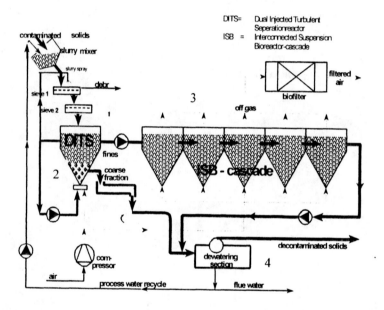

Figure 6: Slurry Decontamination Process (SDP)

The Slurry Decontamination Process has been developed as a (semi)-continuous system (Figure 6). This process contains 4 major unit operations (Oostenbrink *et al.* 1995):

- the contaminated solids are mixed with (process) water into a slurry and sized over a vibrating screen. In this wet sieving step, the debris is removed and a slurry prepared having the required density (25-35 w/w%)

- in the first reactor/separator, a tapered air lifted bioreactor: the DITS-reactor (Luyben and Kleijntjens 1988) the sand fractions are removed by means of a fluidised bed. Extensive organic material is removed by fine screening of light material. In addition, the agglomerates of the contaminated fines are demolished due to the power input and therefore opened to biological breakdown (also inoculation with the active biomass takes place)

- In a second reactor stage the fine fraction is in a cascade of air lifted tapered bioreactors which are connected (ISB-cascade)

- a dewatering stage completes the process, the water released is partly recirculated as process water to mix up the fresh solids into a slurry

At various scales, ranging from mini-plant level (40 l) up to a 4-m³ parts of the system was extensively tested. A pilot plant was operated for two and a half years to test various solid waste streams at a scale of 3 m³. During several treatment periods in which various waste streams was processed, steady states degradation was measured (Kleijntjens and Luyben 1999).

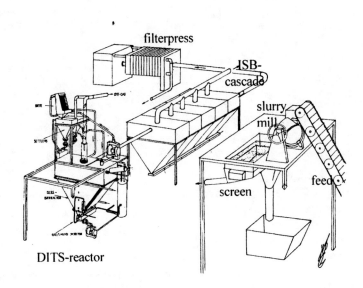

Figure 7: Artist impression of the pilot plant (units not in proper size)

Figure 7 gives an artist impression of the 3 m³ pilot plant, which was constructed by BIRD Engineering in the Netherlands. Shown is how the beginning of the process (the slurry preparation) takes place at a platform while the reactor components are at a lower level. By hydrostatic forces the slurry runs through the system. After the last treatment step, the fifth compartment of the ISB-cascade, the slurry is pumped into a chamber filter press for dewatering.

4.3. MICROBIAL BREAKDOWN IN THE SDP

A contaminated soil treated in the pilot plant resulted in a steady state breakdown pattern as depicted in Figure 8. The measure steady state concentrations (mineral oil), in each of the four units (slurry mill, DITS-reactor, ISB-cascade and dewatering stage), are depicted as a function of time. For a scattered input (black triangles for the input ranging from 600 to 1300 mg/kg), the steady state concentration further on in the process also fluctuated (+/- 350 mg/kg for the DITS-reactor and +/- 200 mg/kg for the ISB-cascade).

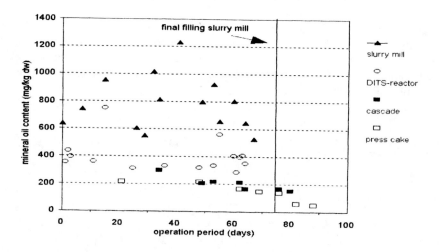

Figure 8: Steady state mineral oil breakdown pattern in the SDP (soil)

After dewatering, end concentrations were reached at levels below 50 mg/kg, well below the recycling standards (residence time was 38 days). Due to scattered input, no smooth steady state breakdown pattern is observed. Each following step, however, clearly has lower steady state values.

Figure 9 shows the experimental results for a heavily polluted sediment (Petroleum harbor-Amsterdam) over a steady state period of 6 weeks, the residence time in this experiment was 16 days. Nutrients (nitrogen, phosphorus and potassium) were added and the temperature was kept at 30 degree Celsius. The steady state PAH concentration in the solids is depicted as a function of time. The upper symbols show the feed concentration in the slurry mill with an average of about 350 mg/kg. In the DITS-reactor the steady state concentration dropped to values around 100 mg/kg (first part of the microbial breakdown).

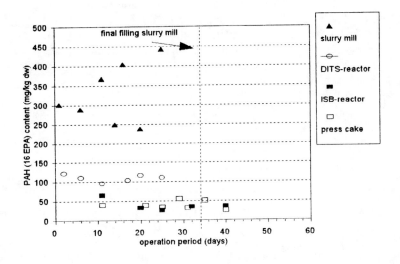

Figure 9: Steady state PAH breakdown pattern in the SDP (sediment)

In the ISB-cascade the average concentration dropped to values of 30/40 mg/kg. After dewatering the final concentration increased somewhat in the filter cake. The overall PAH-degradation level was about 92 % and recycling standards could be reached. No significant evaporation was measured in the off gas.

In this sediment the mineral oil degradation pattern showed a similar trend, however an overall degradation level of only 65 % was established (16 days residence time). Final concentrations after dewatering were around 3000 mg/kg, far above the recycling target of 500 mg/kg. The load of mineral oil to be degraded thus did not match the microbial breakdown capacity generated. It therefore was necessary to reduce the load of contaminants being fed to the system. Since the contaminants predominantly are adsorbed to the soil organics and mineral fines, froth (foam) flotation at the start of the SDP (in the DITS-reactor) was experimentally tested.

4.4. SDP-IMPROVEMENTS: FROTH FLOTATION IN THE DITS-REACTOR

Froth (foam) flotation is a technique used to separate particles if there is sufficient difference in their hydrophobicity. In an aerated particle suspension the hydrophobic particles tend to be collected at the air bubble/water interface; with the air bubbles these particles rise to the surface and gather in the foam on top of the slurry. If the mixture of froth and hydrophobic particles is collected from the top by means of a skimmer, a particle separation is achieved (Perry and Green 1984). In industry flotation is often used to separate metals from a suspension.

Flotation experiments were carried out by connecting a flotation cell to the DITS-reactor. Table 2 gives the mass balances for the DITS-reactor without and with flotation for a sand free sediment. In the standard situation (no flotation) a batch of sediment was fed to the reactor (235 kg) with a solids hold-up of 54%. This resulted in 126.9 kg of solids weight at the start. This load contained 12.18 kg organics (9.8 %) and 2.09 kg of mineral oil (at 17 gr/kg). Without flotation the only separation for this sediment took place at the debris screening, 98% of the solids remained in the DITS-reactor to be treated. The mineral oil load being fed to the biomass was almost (98%) equal to the input.

Table 2: Mass balances for the DITS-reactor

Standard DITS

	flow kg	solids hold-up kg/kg	solids flow kg	organic matter kg/kg	organic flow kg	mineral oil kg/kg	oil flow kg
sediment input	235.000	0.540	126.900	0.096	12.182	0.017	2.094
debris (screen)	5.000	0.480	2.400	0.149	0.358	0.017	0.040
sand flow	0.000	0.000	0.000	0.000	0.000	0.000	0.000
flotation flow	0.000	0.000	0.000	0.000	0.000	0.000	0.000
amount to treat			124.500		11.825		2.054
% of input			98		97		98

DITS+flotation

	flow kg	solids hold-up kg/kg	solids flow kg	organic matter kg/kg	organic flow kg	mineral oil kg/kg	oil flow kg
sediment input	235.000	0.540	126.900	0.096	12.182	0.017	2.094
debris (screen)	9.000	0.480	4.320	0.149	0.644	0.017	0.071
sand flow	0.000	0.000	0.000	0.000	0.000	0.000	0.000
flotation flow	250.000	0.130	32.500	0.230	7.475	0.033	1.073
amount to treat			90.080		4.064		0.950
% of input			71		33		45

Using flotation the mass balance changed significantly. For a load of 126.9 kg solids an amount of 32.5 kg solids due to flotation was separated; the remaining load was 71% in the input. With the flotated solids 7.47 kg of organics and 1.07 kg of mineral oil was removed. Due to flotation only 33% of the organics and 45% of the mineral oil remained to be treated. For PAH the flotation was less effective, 80% of the PAH still were to be treated in the process.

From these experiments it was concluded that flotation is an effective way to reduce the load of contaminants in the feed flow before the bioreactors are entered. Flotation can be used as a "safety measure" to protect the microbial population from overloading. An additional effect of flotation in the DITS-reactor is found in the capability of metal removal with the froth. For waste streams having both organic and metal contaminants, flotation opens the possibility of treating cocktails and physically removing metals without a significant change in the process.

5. Scale-up

Having extensively studied the major process and operational features of the SDP at pilot scale, a feasibility study for a larger demonstration plant has been made. As an example the features of plant with a working volume of 1200 m³ (average capacity of +/- 25,000 tonnes/yr) were determined. At this capacity:

- the hydrodynamics is well known, no significant scale-up problems are to be expected

- the kinetics are expected to be identical to the pilot plant

- market demonstrations can be given

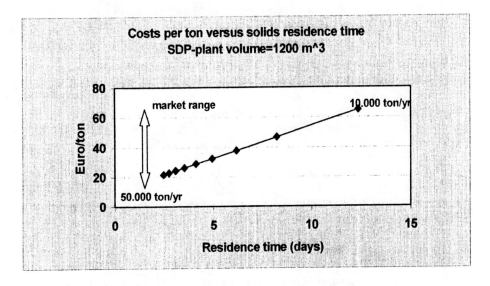

Figure 10: Residence time versus estimated costs per tonne (1200 m³ SDP-installation; cost level=1999)

5.1. PROCESS ECONOMICS

For a defined SDP-plant volume the actual capacity directly is proportional to the solids residence time. The residence time in its turn defines the amount of power needed to process one tonne of solids. The costs per tonne (among costs as labour and finances) thus are a function of the solids residence time. Figure 10 gives the relationship between calculated cost levels per tonne input and the solids residence time. The present Dutch market range (roughly in between 20 and 60 euro/tonne) is depicted by the arrow. Shown are the upper and lower capacities at the example of a 1200 m³ plant volume. At the upper capacity (50,000 tonnes/yr at 2 days residence time) costs are at the lower end of the range; at the lower capacity (10,000 tonne/yr at 12 days residence time) the costs are in the upper range.

Figure 10 is based on the process features such as denoted in Figure 5. This means that as much sand as possible is removed in the DITS-reactor. Flotation effects have not been incorporated in the calculations. After the DITS-reactor, the fines are treated in the ISB-cascade. For the calculations a moderately contaminated soil was taken.

Relating the design results (shown in Figure 10) to the pilot experiments, it has to be concluded that the solids residence times used at pilot scale (ranging from 16 to 38 days) should not be applied. Focusing on the shorter residence times (in between 2 and 12 days) for a scaled up system, two options are available to operate within the economic context:

- flotation in the DITS-reactor

- combination of the SDP with extensive (low cost) treatment of the fines

5.2. EXTENSIVE (LOW COST) TREATMENT OF THE FINES

Extensive treatment of fines is possible in various ways. They all have in common that little handling is needed and that time and natural processes are allowed to improve the quality of the material. The measures to be taken depend on the nature of the contaminants concentrated in the fines. Two major groups may be distinguished being (heavy) metals and organic compounds such as mineral oil, polynuclear hydrocarbons (PAHs), chlorinated hydrocarbons and pesticides. Both groups of contaminants require a different approach.

In the case of heavy metals the environmental risk may be reduced by immobilization or by decontamination. As a result of the treatment process in general a partial mobilization of metals takes place. Natural ripening processes will lead to re-adsorption of metals on clay minerals and the transfer of metals from soluble into insoluble species such as the sulphides. These processes may be obtained by allowing the fines to settle and drain in a basin constructed with earthen walls. Based on the chemical composition of the fines additives may be added to increase the rate of the immobilization processes in which microorganisms may play an important role. These additives will specially composed on the basis of the nature of the project.

Removal of metals may take place by the use of plants (phytoremediation). Various plants can be used, which depends on the nature of the contaminants and the use, which can be made of the products (fire wood, paper, etc.). The choice of the preferred plants is depending of the way in which the plants store the contaminants. The easiest plants are those which collect the contaminants in the leaves. Harvesting, composting, incineration of the residue and collecting the ashes, is in this case the cheapest way to collect the contaminants.

Organic compounds may be removed by microbial processes. Depending on their nature these may be aerobic (mineral oil, PAHs, pesticides, etc.) or anaerobic (chlorinated hydrocarbons, pesticides, etc.). Anaerobic processes are relatively easy to obtain, because the fines will render anaerobic conditions shortly after they settle. Additives are required to fuel, for example, dehalogenation when not present naturally in sufficient amounts and of the right source.

To obtain aerobic degradation, measures have to be taken such as the addition of additional nutrients and oxygen releasing compounds before spreading the fines to the ripening fields. A biological method is the use of plants, which have the properties to transport oxygen to their root zone such as reed (*Phragmitis communis*). In principle an aeration system may be applied, but this can very much intensify the treatment process, which sometimes is not the object. The use of appropriate plants will also improve the soil structure and will increase the rate of dewatering and, thus, decrease the period in which the area is difficult to use.

Above a number of possibilities are indicated for the use of extensive treatment of the fines in order to give these a function in land planning operations for non-food agricultural applications or as a recreational or natural environment. Based on the scale of the operation the result thereof may be more or less extensively reviewed ecologically.

Figure 11 shows the combination of the SDP and an extensive treatment in which the fines are ripened. In this combination after a short residence time in the SDP, the fines are further extensively treated.

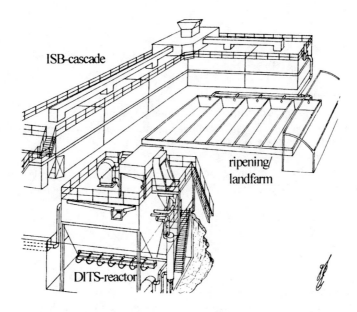

Figure 11: Artist impression of the SDP in combination with extensive after treatment (units shown are not in their proper sizes)

5.3. ENVIRONMENTAL EFFICIENCY OF THE SDP

In order to determine the environmental efficiency of the SDP, firstly the process has to be put into a broad framework (see also Table 1). Being an *ex situ* treatment method, the SDP-process requires an installation using power, in addition, the feed has to be excavated, transported and stored. To operate these type of installations at full scale efficiently, clearly recycling products have to be the outcome of the SDP-treatment. Making recycling products out of waste, however, not only demands a properly designed process, also the logistics of the feed flow, the product flow and quality of the input are crucial. In addition, the economic, legal and political context should be in favour of stimulating the use of recycling products in various parts of the economy.

The overall environmental efficiency for the SDP is determined by the technical features (such as power input per tonne solids and residence time), the quality of the input and output (recycling sand and clay) and the aspects mentioned above.

Having fully explored the SDP-technology and its environmental efficiency, it was concluded that, given the present situation, the basic design of the SDP, using reactor technology only is not appropriate. Instead of focusing on reactor treatment, the combination of reactor operation (short residence times) with extensive methods is to be preferred. In this way the benefits of both approaches are to be combined into integrated systems (Figure 11), resulting in environmentally efficient and flexible solid waste treatment.

6. Conclusions

- A transition in treatment technology from waste decontamination towards waste recycling is taking place.

- In the broad spectrum of technologies, ranging from *in situ* soil remediation towards large-scale *ex situ* treatment installations, the category of constructed natural systems (like landfarming, phytoremediation, lagoon treatment, separation basins) fits in between.

- The Slurry Decontamination Process (SDP), developed and operated up to pilot level, basically was designed as an *ex situ* biotechnological reactor process.

- In the continuously operated SDP-pilot plant mineral oil and PAH contaminated solids can be degraded down to (Dutch) recycling standards giving acceptable end products.

- It was concluded that at the present market conditions, only short residence times (few days) in the SDP-reactors are feasible (costs). At these short reactor residence times the combination of the SDP-reactor technology with an extensive post-treatment (such as "constructed natural systems" in the form of phytoremediation) is aimed at. The basic SDP-set-up therefore is transformed in a combination of technologies.

- To obtain environmentally efficient treatment the question should no longer be how a specific technology can be optimised. The major issue is how to make recycle products using any economical combination of operations, which suits the purpose.

Acknowledgement

The authors would like to thank Dr. Apitz (San Diego) for her contribution.

References

Apitz S.E., Pickwell G.V, Meyer-Schulte K.J., Kirtay V, Douglass E., S ; A slurry biocascade for the enhanced degradation of fuels in soils , Fed. Env. Restoration 3 and waste minimization 2, Conference, New Orleans (1994)

Brady N.C, Nature and properties of soils , Macmillan Publ. Comp., New York (1984)

Building Material Act, Bouwstoffenbesluit bodem- en oppervlaktewaterbescherming , TK 22683, Staatsdrukkerij, the Netherlands (1995)

Cheremisinoff, N.P., Encyclopedia of fluid mechanics , vol 5 : Slurry flow technology, Gulf Publishers, Houston (1986)

Cullinane M.J., Averett D.E., Shafer R.A., Male J.M., Truitt C.L., Bradbury M.R., Contaminated dredged material , Noyes Data Corporation, New Jersey (1990)

EPA, Hazardous Waste Management Division first 5-year review: French Limited Site, Crosby, Texas, CERCLIS TXD-980514814, US (1994)

Harmsen J. Van den Toorn, A., Boels D., Vermeulen B., Ma W., Van der Waarde J., Duin R., Kampf R., Growing Biomass to stimulate bioremediation: technical and economic perspective , In situ and on-site Bioremediation, the fifth International Symposium San Diego (1999)

Hinsenveld M., Shrinking core model as a representative of leaching processes in cement stabilization waste , p 1251, in Proceedings of Contaminated Soil , Brink, Bosman and Arendt (eds) Kluwer, Dordrecht, the Netherlands (1995)

Hinze J.O. Turbulence , McGraw-Hill, New York (1959)

INES, Project bioslib , Stichting Europort Botlek Belangen, Rotterdam, the Netherlands (1997)

Kerr J, Pulford I, Duncan H., Wheeler C, Phytoremediation of heavy metal contaminated sites by fibre crops, in Proceedings of Contaminated Soil , p. 1119,Thomas Telford, London (1998)

Kiehne M., Berghof K., Muller-Kuhrt L., Buchholz R., Mobile Revolving Tubular reactor for continuous microbial soil decontamination , in Proceedings of Contaminated Soil , p. 873, Brink, Bosman and Arendt (eds) Kluwer, Dordrecht, the Netherlands (1995)

Kleijntjens R.H., Biotechnological slurry process for the decontamination of excavated polluted soils , PhD-thesis, TU Delft, the Netherlands (1991)

Laine M. Jorgensen S., Pilot scale composting of chlorophenol-contaminated saw mill soil , in Proceedings of Contaminated Soil , p. 1273, Brink, Bosman and Arendt (eds) Kluwer, Dordrecht, the Netherlands (1995)

Leeuwen van J., Pepels A., Zwakhals J., Production of sand and clay from contaminated sediment Conference on Contaminated Sediments-ICCS, Rotterdam, The Netherlands (1997)

Levenspiel O., Chemical Reaction Engineering , Wiley, New York (1972)

Lotter S., Stregmann R., Heerenklage J., Basic investigation on the optimisation of biological treatment of oil contaminated soils , in Proceedings of Contaminated Soil , p 967, Brink, Bosman and Arendt (eds) Kluwer, Dordrecht, the Netherlands (1990)

Luyben K.Ch.A.M., Kleijntjens R.H., Werkwijze voor het scheiden van vaste stoffen , Dutch patent no 8802728 (1988)

Kleijntjens R., Luyben K., Soil Decontamination, waste gas and potable water preparation vol 11b in Environmental Processes, Wiley, (1999)

Northcliff S., Bannick C., Paetz A, International standardization for soil quality , in Proceedings of Contaminated Soil , Edinburgh, UK (1998)

Oostenbrink I., Kleijntjens R., Mijnbeek G., Kerkhof L., Vetter P., Luijben K., Biotechnological decontamination using a 4 m^3 pilot plant of the Slurry Decontamination Process , in Proceedings of Contaminated Soil , Brink, Bosman and Arendt (eds) Kluwer, the Netherlands (1995)

Perry R.H., Green D., Chemical Engineers Handbook , McGraw-Hill International Book Company New York (1984)

Riser-Roberts E., Remediation of petroleum contaminated soils , Lewis Publishers (CRC-Press), USA (1998)

RIZA-report 97-026, Eindrapport POSW fase 2 (1992-1996) , SDU-Den Haag, The Netherlands ISBN: 90-36950 317 (1997)

Robra K., Somitsch W., Becker J., Jernej J.,Schneider M., Battisti A., Off-site bioremediation of contaminated soil and direct re-utilization of all oil fractions , in Proceedings of Contaminated Soil , Thomas Telford, London (1998)

Schegel H.G., Algemeine Mikrobiologie , Thieme Verlag, Stuttgart (1986)

Schotel F., Roeters P., Groen K., Van der Gun J., Treatment of contaminated dredges material by immobilization , International Conference on Contaminated Sediments, p 1159, Rotterdam, The Netherlands (1997)

Staatsuitgeverij, Dutch Building Material Decree, Den Haag, the Netherlands (1995)

Suzuki M., Waste management according to Japanese experience , Fourth World Congress of Chem. Eng., Dechema, Frankfurt, Germany (1992)

Staps J.J.M., International evaluation of *in situ* biorestoration of contaminated soil and groundwater RIVM-report no 738708006, Bilthoven, The Netherlands (1990)

Warbout J. and Ouboter P.S.H., Een vergelijking tussen verschillende methodes om het gehalte aan minerale olie in grond te bepalen , H₂O (the Netherlands), (21), nr 1 (1988)

PART 3
IN SITU CLEAN UP TECHNOLOGIES

IN SITU BIOLOGICAL SOIL REMEDIATION TECHNIQUES

PETER MIDDELDORP, ALETTE LANGENHOFF, JAN GERRITSE AND HUUB RIJNAARTS

TNO Environment, Energy and Process Innovation
Dept. of Environmental Biotechnology
P.O. Box 342, 7300 AH Apeldoorn, The Netherlands
E-mail: P.Middeldorp@mep.tno.nl

1. Introduction

The *in situ* application of biological degradation processes of soil pollutants on a field scale is approximately 10-15 years old. Numerous laboratory and mesocosm studies have been published to form a sound basis for the application of these techniques. At present, *in situ* biological soil clean up has evolved to a full-fledged and cost-efficient alternative to other remediation techniques.

During the last decade, many biological degradation processes of various pollutants in different soil types have been developed into *in situ* bioremediation techniques. Among these techniques, we also classify spontaneous biodegradation processes in soil (natural attenuation), provided that it is adequately monitored for acceptable prolongation of this process in time.

The approach to *in situ* bioremediation of industrially contaminated sites depends both on the geohydrological and spatial situation of the site and on the type of contaminant in relation to the biodegradation processes that may take place in the soil.

The first aspect can be viewed from Figure 1. The contaminated area normally consists of:

- A source zone. Mostly, high pollutant concentrations in the source zone will be removed using non-biological techniques, although innovative (combined) biological techniques are also used.

- A plume zone. Compared to the source zone, this zone normally contains much lower pollutant concentrations, which can be used by degrading microorganisms. Here, we sometimes have to enhance the biodegrading activity by creating the appropriate conditions (active plume management). When conditions for biodegradation already prevail, we speak of:

S.N. Agathos and W. Reineke (eds.),
Biotechnology for the Environment: Soil Remediation, 73-90.
© 2002 Kluwer Academic Publishers. Printed in the Netherlands.

- A natural attenuation zone. Here, provided that there is sufficient spontaneous biodegradation, the plume will possibly not proliferate any further. Obviously, natural attenuation processes may also take place in the source and plume zones.

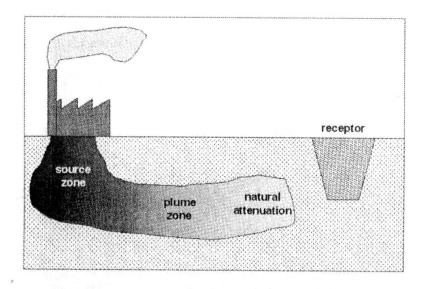

Figure 1: Schematic overview of a typical industrially polluted site and the different zones, which determine the type of bioremediation technique. These different zones and the specific in situ bioremediation techniques will be described in this chapter.

The second aspect is the biodegradability of the pollutant under the environmental conditions that prevail or may be created in the soil. Figure 2 shows a typical picture of biodegradation rates of different types of pollutants that may occur under different redox conditions. It is shown that BTEX, mineral oil, PAHs (polycyclic aromatic hydrocarbons) and lower chlorinated compounds can be readily mineralised under aerobic conditions. However, the lower the redox potential, the lower the relative biodegradation rate becomes.

Highly chlorinated compounds (e.g. hexachlorobenzene, tetrachloroethene) are essentially not degradable at high redox potentials. Such compounds can be dechlorinated to lower chlorinated analogues under methanogenic and sulphate-reducing conditions. Therefore, to achieve total mineralisation of these compounds, sequential anaerobic and aerobic conditions are often required.

When the aforementioned spatial and quantitative distribution of the pollutants in the soil and the prevailing local redox conditions are considered, one can assess whether or not *in situ* bioremediation is feasible and in which form it can be applied. Without pretending to be complete, Table 1 shows a number of soil pollution cases to which different *in situ* biological clean up techniques are being or have been applied. In the following, we will elaborate on some of these cases, as examples of the aforementioned considerations and the resulting actions to be taken. The discussion will be essentially on relatively mobile contaminants, such as BTEXN (benzene, toluene, ethylbenzene, xylenes and naphthalene) and VOCl (volatile organic chlorinated compounds, e.g. chlorinated ethenes, ethanes and methanes).

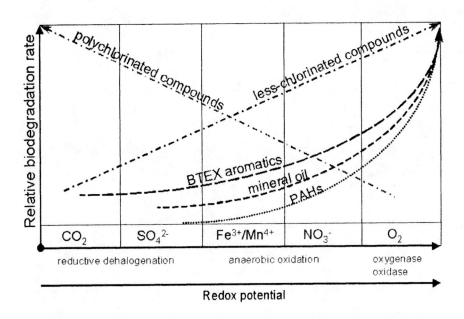

Figure 2: *Biodegradation rates and dominant degradation mechanisms for different classes of pollutants under various redox conditions.*

Finally, to place *in situ* soil clean up in a larger context, the management of large contaminated areas will shortly be discussed. These so-called "mega-sites", for which just the use of a number of different remediation techniques is not sufficient, need to be viewed in the full socio-economical context before decisions on clean up or pollution control can be made.

2. Source zone remediation techniques

The source zone of a contamination is typically the place where the pollutants have entered the soil, and is characterised by the presence of high concentrations or even pure phases of the pollutants. The necessity of cleaning up this zone is obvious, as it acts as a long lasting source of continuous delivery of the pollutants to the surrounding soil compartments.

Table 1: In situ soil bioremediation at field scale

Site	Soil type	Pollutant(s)	Source / plume	Remediation strategy	E-donor / acceptor / substrate	Results (contaminant reduction)	Ref. *
Groningen, The Netherlands		PCE, TCE	Source	Anaerobic biostimulation	Methanol, compost leachate	PCE, 95% (ethene/ethane produced)	1
Pernis, The Netherlands	Sand	BTEX, min. oil	Plume	Air sparging	O_2	All contam. 50%	2
Hengelo, The Netherlands		HCH, MCB, benzene	Plume	Anaerobic biostimulation	Methanol / O_2	Not yet available	3
DuPont Plant West Landfill, TX	Sand	PCE, TCE	Un- known	Anaerobic biostimulation	Benzoate / sulphate	PCE, TCE, DCE <5 ppb, VC <10 ppb	4
Breda, The Netherlands	Sand	PCE, TCE, DCE	Source + plume	Anaerobic bio- stim.+aerobic co-metabolism	Methanol / phenol	>98% all contaminants (2 yrs.)	6
St. Joseph, MI, transect to lake	Un- known	TCE, DCE, 11DCA, VC	Un- known	Natural attenuation	COD / CO_2, SO_4, Fe^{3+}, O_2	>99% all contaminants	7,8
Tilburg, The Netherlands	Sandy silt	PCE, TCE, DCE, BTEX, min. oil	Plume	Stimulated nat. attenuation by plume mixing	BTEX/ PCE, TCE, DCE, VC	PCE, TCE >98%, DCE ca. 50%	9
Plattsburgh Air Force Base, NY	Un- known	TCE, DCE, VC	Un- known	Natural attenuation	BTEX / SO_4, Fe^{3+}, O_2	TCE, DCE, VC>99%, BTEX no data	10
Hemingway, SC	Silty clay	MTBE, BTEX, naphthalene	Un- known	Air sparging + soil vapour extraction	O_2	MTBE, BTEX >99%, naphthalene 96%	11
Miamisburg, OH	Sandy gravel	PCE, TCE, DCE, VC, CF, TEX	Source + plume	Air sparging + soil vapour extraction	O_2	VOC in saturated zone down from 618 to 5 ppm	11
Chattanooga, TN	Clay	MTBE, BTEX, TPH	Plume	Infiltration of oxygenated water + bio- augmentation	O_2	MTBE 96%, benzene 88%, TPH 82%	11
Ravenel, SC	Un- known	MTBE, BTEX, naphthalene	Source	Excavation + injection of ORC^x	O_2	MTBE and naphthalene < det. limit, BTEX 95-99%	11
Drycleaner, Orlando, FL	Sand	PCE, TCE, DCE, VC	Plume	Hydrogen releasing compound (HRC)	H_2	PCE 96%, TCE 51%, *cis*-DCE 36%, VC 58%	11
Hilton Head, SC	Un- known	MTBE, BTEX, naphthalene	Source	Bioaugmentati on surfactant + encapsulant	O_2	MTBE 96%, benzene 83%, toluene 66%, naphth. 84%	11

* 1. (Langenhoff *et al.* 2001a) 5.(Semprini *et al.* 1992) 9. (Smittenberg *et al.* 2000)
  2. (Langenhoff *et al.* 2001b) 6.(Spuij *et al.* 1997) 10. (Wiedemeier *et al.* 1996)
  3. (Langenhoff *et al.* 1999) 7.(Adriaens *et al.* 1997) 11. (USEPA 2001)
  4. (Beeman *et al.* 1994) 8.(McCarty and Wilson 1992)

Depending on the size of the source zone and the depth of the contamination, it may be economically favourable to excavate the soil and subsequent disposal. Most *in situ* techniques for source removal are based on physical-chemical techniques, such as steam injection, surfactant enhanced cleaning, co-solvent flushing, etc. However, depending on the concentration of the contaminants, there are also biological remediation techniques available, which can be applied alone or in combination with physical-chemical techniques.

2.1. REDUCTIVE DECHLORINATION AT THE RADEMARKT, GRONINGEN, THE NETHERLANDS

2.1.1. Introduction

The Rademarkt site in the centre of the city of Groningen, The Netherlands, is contaminated with tetrachloroethene (PCE) and trichloroethene (TCE). Mixed redox conditions control intrinsic biodegradation processes and intrinsic degradation had taken place at this site, as was evidenced by the detection of degradation products (*cis*-dichloroethene (*cis*-DCE) and vinyl chloride (VC)). This intrinsic reductive dechlorination may have been enhanced by an old -temporary- leaking sewage system, providing the electron donors that are needed for the dechlorination processes (de Bruin *et al.* 1992, Middeldorp *et al.* 1999). However, the natural transformation rates of *cis*-DCE and VC at the site were too low to prevent migration of these hazardous compounds from source zone to areas that must be protected. The low levels of DOC in the groundwater (< 10 mg/l) indicated a lack of naturally present electron donor and addition of electron donor was needed to achieve a complete degradation of the chlorinated ethenes at the site (de Bruin *et al.* 1992, Middeldorp *et al.* 1999).

The potential of the dechlorination processes at the Rademarkt site was previously demonstrated in batch- and column experiments (Nipshagen *et al.* 1999, Van Aalst-Van Leeuwen *et al.* 1997). Based on these experiments, a full-scale system was designed to treat the source of the contamination by enhancing the reductive dechlorination with both methanol and compost leachate as electron donors. The system included an infiltration and recirculation system and was implemented at the site.

2.1.2. Pilot study

The pilot system at the Rademarkt site (65 x 20 m) consisted of 10 infiltration wells on one side and 5 extraction wells at the other side of the source zone at a depth of 5 to 8 m bgl (below ground level). Part of the extracted groundwater was re-infiltrated via the infiltration wells. By recirculating part of the extracted groundwater without purification, the soil was used as a 'bioreactor'. A mixture of electron donor and nutrients was added to the untreated re-infiltrated groundwater. Finally, the infiltrated groundwater contained methanol, compost leachate and ammonium chloride. The pilot test lasted for 35 weeks, and was continued afterwards due to its success.

2.1.3. Results

A few months after start up, nitrate was depleted, sulphate concentration decreased significantly, and methane concentrations increased. This indicates that the soil had become more reduced (sulphate reducing or methanogenic conditions), which is beneficial for the reductive dechlorination process. Increased levels of H_2 and decreasing redox-potentials confirm these results.

The percentage of dechlorination was calculated from the monthly monitoring of chloroethene concentrations, and is expressed as the amount of chlorine that is eliminated from the source compound PCE.

$$Dechlorination\% = \frac{\frac{1}{4}[TCE] + \frac{2}{4}[\sum DCEs] + \frac{3}{4}[VC] + [Ethene] + [Ethane]}{[PCE] + [TCE] + [\sum DCEs] + [VC] + [Ethene] + [Ethane]} * 100\%$$

Dechlorination was established in most of the stimulated area. Starting at an average of 35%, the dechlorination increased to 50 - 95% in the different monitoring wells. A high degree of dechlorination (> 75%) was found primarily in the northern part of the stimulated area and in close proximity to the infiltration wells in the southern part. The concentrations of PCE and TCE decreased over time, and completely dechlorinated products like ethene and ethane were formed. Figure 3 shows an example of a monitoring well in which the sequence of dechlorination products was evident, and the dechlorinated end products, ethene and ethane, were produced.

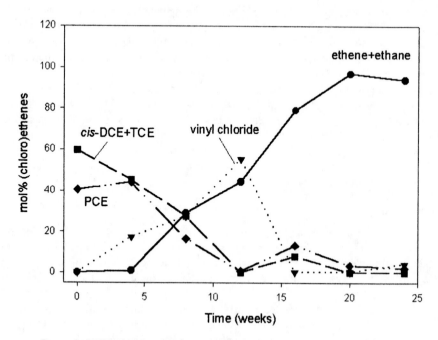

Figure 3: Mol% of (chloro)ethene concentrations in a monitoring well. The total initial concentration of chloroethenes was around 10 μM.

The central part of the contaminated site showed less degradation, as the added electron donor was not able to reach this area. This was due to malfunctioning infiltration filters: changes in infiltration strategy will solve this problem in the future.

Detection of chloroethene-dechlorinating *Desulfitobacterium* spp. at the site indicated increased numbers during the infiltration processes.

The successful performance of this pilot was continued: the infiltration system is still in use to remediate the site.

2.2. AEROBIC *IN SITU* BIOREMEDIATION TECHNIQUES

Aerobically biodegradable constituents like BTEX, PAH, mineral oil etc. can theoretically be treated by aerobic *in situ* treatments, like air sparging and bioventing. Air sparging is an *in situ* remedial technology that reduces concentrations of volatile compounds that are dissolved in groundwater. This technology involves the injection of air into the subsurface saturated zone, enabling a phase transfer of hydrocarbons from a dissolved state to a vapour phase. The air is then vented through the unsaturated zone.

Air sparging has been found to be effective in reducing concentrations of volatile organic compounds found in petroleum products at underground storage tank sites (Alleman and Leeson 1999, Leeson *et al.* 2001). Air sparging is applicable for the lighter gasoline constituents (*i.e.* BTEX), because they readily transfer from the dissolved to the gaseous phase. Air sparging is less applicable to diesel fuel and kerosene.

Bioventing is an *in situ* remediation technology that uses indigenous microorganisms to biodegrade organic constituents adsorbed to soils in the unsaturated zone. The activity of the indigenous bacteria is enhanced by inducing air (or oxygen) flow into the unsaturated zone. This is done by using extraction or injection wells. Bioventing has proven to be very effective in remediating releases of petroleum products including gasoline, jet fuels, kerosene, and diesel fuel. Bioventing is most often used at sites with mid-weight petroleum products *i.e.* diesel fuel and jet fuel (Alleman and Leeson 1999, Leeson *et al.* 2001).

3. Active plume management techniques

To which extent a soil pollution plume will develop depends mainly on the mobility of the pollutant, the time the source has been there, the groundwater velocity, the geohydrology of the site and the intrinsic biodegradation rate. These factors will determine whether a plume will reach a defined object at risk (receptor) or that natural attenuation will be capable of preventing this. Often, it will be decided that active plume management in the form of enhanced biodegradation is necessary to protect the receptor. As the goal of such an intervention is to prevent further spreading of the plume, it is sufficient to create a zone with enhanced biodegrading activity, perpendicularly to the direction of the plume movement.

Depending on the situation, the biodegrading activity can be enhanced by the creation of appropriate redox conditions, addition of nutrients, electron donors and eventually by the addition of degrading microorganisms (bio-augmentation).

3.1. HCH REMOVAL AT AN INDUSTRIAL LOCATION IN THE NETHERLANDS

At an industrial site in The Netherlands, high concentrations of different HCH isomers have been found in the groundwater. A large HCH containing groundwater plume was moving towards an adjacent canal. To prevent contamination of the canal, HCH had to be removed from the groundwater before the plume reached the canal. In contrast to other HCH isomers, β-HCH is known to be recalcitrant to biodegradation under aerobic conditions. However under anaerobic conditions, all isomers, including β-HCH, can be microbiologically transformed to the intermediates monochlorobenzene (CB) and benzene (Figure 4) (Middeldorp *et al.* 1996, Van Eekert *et al.* 1998). As these intermediates can be further degraded under aerobic conditions, complete mineralisation of all HCH isomers is possible (Reineke and Knackmuss 1984).

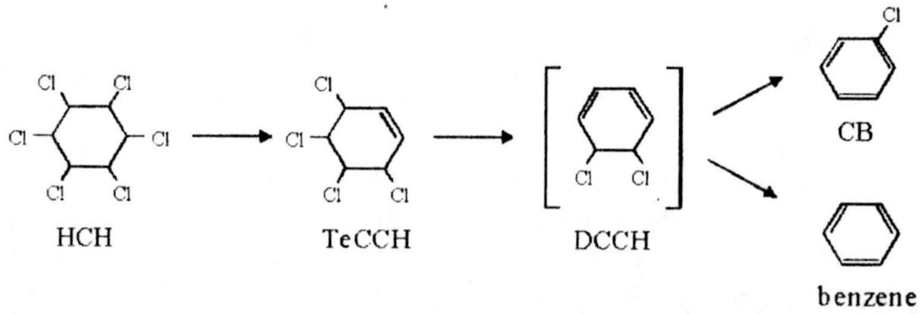

Figure 4: Anaerobic biodegradation pathway of HCH. HCH is reductively dechlorinated to tetrachlorocyclohexene (TeCCH), dichlorocyclohexadiene (DCCH) to form the end products benzene and chlorobenzene (CB). Scheme from (Middeldorp et al. 1996)

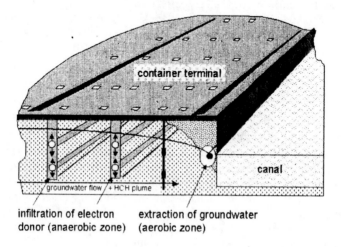

Figure 5: Cross-section of a sequential anaerobic-aerobic biological screen with electron donor infiltration and extraction drain facilities to treat the HCH plume

Characterisation of the industrial site indicated that natural attenuation occurred, but that the rate was too low to prevent infiltration of HCH into the canal (Figure 5). Therefore, an *in situ* biological method for HCH-contaminated soils and groundwater has been developed, which uses bioscreens/activated zones and natural bioremediation processes. As part of the redevelopment of the site, infiltration facilities have been installed to create an anaerobic electron donor infiltration zone in which HCH transformation into CB and benzene will be stimulated. Since aerobic conditions are required for the subsequent degradation of CB and benzene, the anaerobic infiltration zone is followed by an aerobic step. The specific design of the system has been determined by the combination of a bioremediation system, and the redevelopment of the site as a container terminal control (Figure 5). The first results of the groundwater analysis indicate an increased amount of benzene and monochlorobenzene in the groundwater.

3.2. MINERAL OIL AND BTEX REMOVAL AT A HARBOUR SITE IN THE NETHERLANDS

3.2.1. Introduction

In a harbour area in The Netherlands, a contaminated groundwater run-off towards the harbour surface water had to be prevented. The contaminants consisted mainly of BTEX and mineral oil. Therefore, a specific pilot site was selected for the construction of three aerobic biological fences. The pilots are based on air injection in soil at a depth of 4 m-bgl. Two fences consisted of horizontal infiltration drains in either the original soil material (Fence 1) or in gravel (Fence 2). The third fence had vertical infiltration filters that were installed in the original soil for the injection of air. The fences were tested for their capacity to remove aromatic compounds and mineral oil from the groundwater.

3.2.2. Performance of the fences

The biological fences were operated for two years. After starting the air injection, oxygen was detected in the monitoring wells, and a moderate increase in the redox potential was found. This indicates that more oxidised conditions (favourable for hydrocarbon degradation) were established.

Analysis of the contaminants demonstrated that the fences were biologically active, and that the concentrations of hydrocarbons in the fences decreased over time. The concentrations of total hydrocarbons decreased in fence 1 from 17 000 to 9 000 µg/l, in fence 2 from 3 500 to 1 600 µg/l and in fence 3 from 2 000 to 1 000 µg/l (Figure 6).

Fence 1 (horizontal infiltration drain) showed the largest decrease in concentration, but this was also the fence with the highest initial contaminant concentration. When calculating the decrease in percentage, the three fences acted similarly and showed 50 % reduction of the contaminants.

3.2.3. Technical comparison

The flexibility with respect to construction and air injection is high at fences with vertical filters. The flexibility of fences with horizontal injection drains is low. Based on

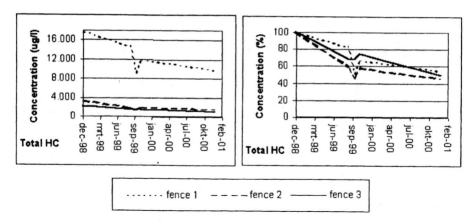

Figure 6: Performance of the aerobic fences

the technical comparison, the fence with vertical filters is the most cost-effective technique for air injection, both from construction and operation point of view. However, in the presence of buildings and infrastructure, horizontal drains are the only alternative. The costs of biological fences are comparable to conventional systems at the site (pump and treat). However, biological fences have less environmental drawbacks than conventional techniques.

4. Natural attenuation

Natural attenuation is defined as the loss of contaminant mass and concentration over time and space in a contaminant plume. Monitored natural attenuation (MNA) is becoming more and more accepted as remediation approach (Rittmann 2000). An important reason is the recognition that engineered remediation is often too expensive or ineffective, especially for very large and deep plumes and for impermeable, clay/loam containing aquifers. Intrinsic biodegradation, mediated by indigenous microorganisms, is the most important natural attenuation mechanism. It can indeed lead to complete removal of pollutants from groundwater systems. The improved understanding of microbial metabolism of pollutants, and the environmental factors that affect this process in the subsurface, support an increasing confidence to apply MNA. To use MNA as plume management approach, clean-up strategies should be considered with site owners and the authorities. Three types of contaminant plumes can be distinguished:

- Expanding plumes: the rate of contaminant transport exceeds the rate of natural attenuation;

- Stationary plumes: the rate of contaminant transport is matched by the rate of natural attenuation;

- Shrinking plumes: the rate of natural attenuation exceeds the rate of contaminant transport.

Dutch policy requirements state that a stationary or shrinking plume has to be achieved within 30 years. To comprehend and predict plume behaviour, the following information should be obtained.

- Evidence for biodegradation of the pollutants in the subsurface;

- Assessment of the extent and rate of in situ biodegradation;

- Prediction of plume behaviour, indicating whether the plume will: i) expand, ii) stay or become stable, iii) stay or tend to shrink, iv) be remediated within a specific timeframe;

- Evaluation of the current and future use of the contaminated site, including potential risks for site users and receptors such as drinking water wells;

- Data collection from a network of wells to monitor pollutants and biological activity over an extended period of time.

4.1. INDICATORS FOR *IN SITU* BIODEGRADATION

To verify that a (bio)degradation takes place, *in situ*, groundwater samples can be used to detect various indicators of biological activity. Often, specific metabolites are formed during the degradation of pollutants. Well-known examples are *cis*-DCE and VC from anaerobic PCE and TCE transformation, chloroethene-epoxides from aerobic co metabolism of TCE, *cis*-DCE and VC, benzene and chlorobenzene from anaerobic HCH transformation, chlorobenzoates from aerobic PCBs metabolism, and lower chlorinated benzenes from anaerobic hexachlorobenzene dechlorination.

Degradation of significant pollutant concentrations is also reflected in the redox chemistry of the groundwater because it is coupled to the reduction of O_2 and other electron acceptors available in the subsurface. Sequential terminal electron accepting processes (TEAPS) often dominate throughout a contaminant plume and can be detected as electron donor/acceptor gradients. Many plumes have a methanogenic core, surrounded by sulphate-, iron-, nitrate- and oxygen-reducing zones. The dissolved H_2 concentration in the groundwater is a good indicator of the dominating TEAP. The sequential redox conditions can be important for the complete degradation of the contaminants (Figure 2). For instance, reductive dechlorination processes may proliferate in the reduced (methanogenic) core, where fermentation products and H_2 are the electron donors. Although in the core, the sediment fraction may release electron acceptors like Fe^{3+} and Mn^{4+}, the oxidation of products like DCE, VC, BTEX or petroleum hydrocarbons mainly occurs at the plume fringes. Here, electron acceptors become available through mixing of contaminants with nitrate and oxygen-containing groundwater.

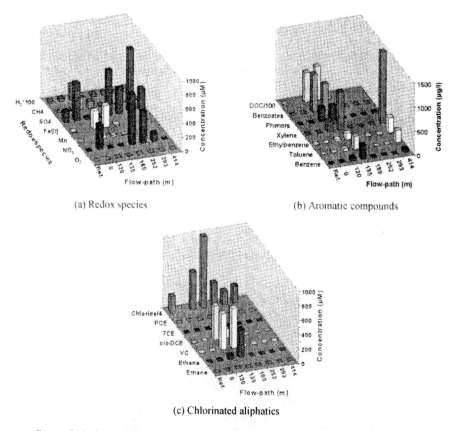

(a) Redox species

(b) Aromatic compounds

(c) Chlorinated aliphatics

Figure 7: Redox-species (a). aromatic compounds (b) and chlorinated aliphatics (c) in monitoring wells along a groundwater flow-path on a contaminated train revision site. Distance from the source is indicated on the X-axes in metres. Ref.: non-contaminated reference monitoring well. Data from (Van Liere et al. 1999)

An interesting example of various indicators for *in situ* biodegradation was found on a train revision site in The Netherlands (Van Liere *et al.* 1999). In the subsurface, different source zones of BTEX, oil and chlorinated solvents were detected (Figure 7). In the wells with BTEX and oil contamination, high concentrations of degradation products were found. The detection of methyl- and ethylphenols and benzoic acid derivatives indicated the degradation of aromatic contaminants, whereas the presence of C_2-C_7 fatty acids pointed to the metabolism of alkanes. Elevated NO_2^-, Mn^{2+}, Fe^{2+}, HS^-, and CH_4 concentrations in the groundwater revealed the reduction of NO_3^-, Mn^{4+}, Fe^{3+}, SO_4^{2-}, and CO_2. Hydrocarbon degradation also generated an increased dissolved H_2 concentration, which in combination with the organic acids, likely acted as an electron donor for the reduction of PCE. This is suggested by the high DCE, VC, ethene and chloride concentrations in the zones with combined presence of BTEX/oil and chlorinated solvents.

Also, a combination of different microbial analyses can be used to obtain an impression of the relative sizes and heterogeneity of the bacterial populations throughout the redox gradients within a contaminated plume. Laboratory microcosm studies with soil and groundwater from a contaminated plume are useful tools give insight into the presence of populations of microorganisms with different pollutant degradation potential under the various redox conditions anticipated in the field.

The culturable fraction of specific functional bacterial groups - e.g. methanogens, sulphate-reducers, iron-reducers, dechlorinators, BTEX-degraders – can be determined with most probable number (MPN) counts in selective anaerobic media.

A more advanced method to reveal the distribution and composition of the microbial populations in the field is through a molecular approach, targeted at the 16S rDNA genes. After extraction of the DNA of the total microbial population in groundwater or soil samples, the 16S rDNA can be used to identify the dominant bacteria at different locations in the plume. In addition, more specific detection can be used to identify known biodegrading microbes, e.g. dechlorinating bacteria.

The determination of changes in stable isotope ratios of the contaminants and their degradation products may become an attractive method to quantify the extent of *in situ* degradation. Different microbial and abiotic degradation mechanisms have different isotopic fractionation effects. Thus, the appearance of changes in the ratio of e.g. $^{13}C/^{12}C$, $^{2}H/^{1}H$, and $^{37}Cl/^{35}Cl$ isotopes of for example chlorinated solvents and BTEX within a contaminant plume are indicative for biodegradation. For the calculation of *in situ* degradation rates it is necessary to know the initial isotope ratios of the spilled contaminant, and to obtain "fractionation coefficients" from studies in laboratory cultures (Morasch *et al.* 2000).

4.2. NATURAL ATTENUATION RATES AND PLUME BEHAVIOUR

Natural attenuation of many pollutants has been investigated at numerous contaminated sites. A review of *in situ* biodegradation rate constants for fuel hydrocarbons and chlorinated solvents indicated that the mechanisms are site-specific and depend on the prevailing biological and geochemical conditions (Suarez and Rifai 1999). Assessment of actual *in situ* biological activity and natural attenuation rates is difficult. The most obvious method is to determine the decrease of the contaminant concentration over time and space. Ideally, a 2 or 3-dimensional GIS-map is constructed of the distribution of the contaminants in the subsurface. Based on a detailed knowledge of the geohydrology and historical data, a numerical model (e.g. MODFLOW, RT3D, MT3D, PHT3D, BIOCHLOR, BIOSCREEN) can then be used to describe plume behaviour with and without the effect of natural attenuation. After that, it is often assumed that the obtained natural attenuation rates will remain constant to make a prediction of the plume status in the future. The models can also be used to simulate the effect of remediation procedures, on the expansion or reduction of contaminant plumes. Source removal, followed by natural attenuation of the plume appears to be an attractive approach to reduce risk and control contaminant spreading on a reasonable cost basis. However, one should be aware that source remediation procedures might actually disturb intrinsic biodegradation processes in the plume. For example, air sparging to stimulate aerobic degradation and to physically remove volatile compounds can increase spreading of contaminants dramatically and disturb anaerobic reductive dechlorination processes

(Norris 2001). Similarly, electron donor supply, to enhance reductive dechlorination, can inhibit degradation of BTEX and petroleum hydrocarbons due to depletion of suitable electron acceptors.

4.3. DIFFICULTIES AND COMPLICATING FACTORS

Although intrinsic biodegradation appears to occur on virtually all sites, natural attenuation is sometimes not effective. This may be due to very small or zero degradation rates. Likely causes may be low concentration thresholds for biodegradation and unavailability of contaminants due to sorption to soil particles or non-aqueous-phase liquid (NAPL) formation. In addition, toxicity of pollutants and degradation products, lack of nutrients (e.g. N, P) may play a role. The xenobiotic character of contaminants may be a reason why microbes with specific biodegradative capacity are not (yet) present on a location. Such a lack of microbes can sometimes be abolished by augmentation with specific microbial populations. Examples are the supply of *Dehalococcoides ethenogenes* to induce *cis*-DCE and VC dechlorination, or the addition of bacteria capable of benzene degradation under nitrate-reducing conditions (Hendrickson *et al.* 2001).

4.4. PLUME MANAGEMENT THROUGH MONITORED NATURAL ATTENUATION (MNA)

In Europe the experience with MNA as plume management tool is still limited. The possibility to apply MNA was investigated at a number of contaminated US Air Force locations. Natural attenuation of BTEX and petroleum hydrocarbons occurred at all 42 sites that were studied (USEPA 2001). Most (35) of the contamination plumes appeared stable, 6 plumes were shrinking and only 1 plume was characterised as expanding. Generally, on the sites containing free-phase products, an average of 30 years would be required to reduce the BTEX concentration to below the cleanup target. Combined with source removal, the remediation times were predicted to drop to less than 20 years. Based on these studies the regulatory authorities approved MNA as plume management approach at 17 of the 42 sites. Natural attenuation of chlorinated solvents took place at all of the 14 contaminated US Air Force sites that were also studied. The plumes were characterised to be expanding (3 sites), stable (6 sites) or shrinking (5 sites). More experience and future plume monitoring will prove whether our current understanding of natural attenuation is accurate and our quantitative prediction of the behaviour of pollution plumes is acceptable to apply MNA as a risk based groundwater remediation strategy.

5. Discussion and outlook

In this chapter we have shown that *in situ* bioremediation is a valuable tool to deal with soil and groundwater pollution, which can be applied both in source and plume zones. If the environmental conditions are suitable, monitored natural attenuation is a responsible and cost-effective plume management approach. We have mentioned chemical and biological parameters that indicate the occurrence of biodegradation of specific compounds, which is not only a useful instrument in natural attenuation approaches, but also when biodegradation is stimulated. Whether or not a biodegradation process has to be stimulated, the key issue remains to understand the biological and physico-chemical processes that take place in the soil aquifer. Such understanding may lead to new tools. For example, in the near future it will probably be possible to detect *in situ* microbial activity by revealing the presence of messenger RNA of specific key-enzymes of biodegradation processes. DNA probes have already been used successfully for detection of aromatic dioxygenases and reductive dehalogenases in laboratory cultures. Improved procedures to extract mRNA from groundwater and soil will make it possible to reveal specific *in situ* activities.

Biological and physico-chemical parameters are used in contaminant transport modelling to predict a.o. the duration of a remediation. Obviously, the modelling approach goes with a relatively large uncertainty, due to the wide time-span of the predictions -in particular in the case of natural attenuation-, the heterogeneity of the subsurface, and the complexity of the various interacting natural attenuation processes. Mass-balances of the donors and acceptors available subsurface are helpful to assess the sustainability of specific reductive and oxidative biodegradation processes. It is likely that the *in situ* degradation rates will not be constant over time. This makes it very difficult to predict long-term natural attenuation processes and the restoration of groundwater systems.

Different guidelines and quick-scans are available to evaluate natural attenuation of BTEX and chlorinated solvents for a range of scenarios. Still, there is a great need for a toolbox with robust long-term tools for the forecasting of natural attenuation. More accurate tools can be established when there is a greater understanding of the underlying processes of natural attenuation such as biodegradation, chemical reactions, sorption, volatilisation and dispersion.

Besides a refinement of the *in situ* remediation methods, our highly industrialised societies and concomitant integrated environmental legislation have forced us to look at soil and groundwater pollution at a different abstraction level. Throughout the world, large areas and regions exist with a high density of industry and polluted sites (megasites, table 2). Examples are seaports, (former) complexes of chemical industry, and mining and military areas. The typical pollution profile of megasites consists of: different classes of pollutants, multiple sources, often located near and under industrial production facilities, and/or large sources with dimensions expanding several square kilometres.

Complete cleanup within an intermediate (25 years) timeframe is in most cases technically and economically not feasible. Therefore, megasites represent steady and long-term potential sources for contamination of groundwater, surface water and sediments in river basins. Besides old infrastructure, the pollution profile has often an inhibitory effect on socio-economic activities. On the other hand, megasites are in many cases key regions for current and new industrial activities and represent opportunities for economic (re)development.

Table 2: European megasites and their basic characteristics.

Megasite	River basin position	Type of Industry	Type of pollution	Size (km^2)
Katowice, Poland	Spring	Mining, metallurgy	Toxic metals	20
Ziar, Slovakia	Spring	Mining metallurgy	Toxic metals, cyanide	500
Black triangle, Czech Republic	Spring, middle part	Mining, lignite, organic chemistry	PAH, toxic metal, halo-organics, cyanide	>1000
Baja Mara, Romania	Middle part	Mining, Metallurgy	Toxic metals, cyanide	>1000
Donana basin, Romania	Middle part	Mining	Cyanide	
Bitterfeld-Wolfen region, one of 22 megasites in Eastern Germany	Middle part	Organic chemistry	Halo-organics	250
Krivoy Rig, Ukraine	Middle part	Mining, Organic Chemistry, metallurgy, Coal	Toxic metals, halo-organics, radio nuclides	>1000
Central Bulgaria	Middle part	Mining	Toxic metals, radionuclides	> 250
Mining sites, Eastern Germany	Middle part	Mining	Toxic metals, radionuclides	> 1000
Rotterdam Seaport	Delta	Min. oil/Petrochemistry	Petrochemicals	1000
Baku, Azerbaijan	Delta	Mineral oil production	Mineral Oil	>1000
Estonia/Latvia/Lithuania	Delta	Military Bases	Ammunition, propellants	not applicable
EU-region	Estimated number and type of megasites			
Current EU states	10.000 - 100.000	operational and some former industrial complexes		
Accession states	1.000 - 10.000	heavily contaminated (former) industrial complexes		
Eastern Europe and Russia	> 10.000	heavily contaminated (former) industrial complexes		

Risk-based and cost-efficient land, groundwater and surface water quality management (also including sediments) is an essential part in stimulating/revitalising the economy and ecology in these regions. Natural attenuation and *in situ* (bio)technologies, included in megasite management approaches, can contribute to this to a great extent. These technologies have been applied to smaller sites, as has been described in previous paragraphs. For megasite application these technologies need to be scaled-up to a regional scale. In many countries, integrated land and water quality management is being initiated (Anonymous 2001). In Europe, compliance with the new EU Water Framework Directive (European Directive 2000/60/EC) enforcing River Basin water quality management is becoming an important driver for this. Similar developments are taking place in Japan and North and South America.

Integrated, large-scale application of natural attenuation and *in situ* biotechnology is an important means to meet such legislative requirements throughout the world and provide a safe and sustainable quality of life and of our environment.

References

Adriaens P, Lendvay J, McCormic ML and Dean SM (1997) Biogeochemistry and dechlorination potential at the St. Joseph Aquifer-Lake Michigan interface. Pages 173-178 in B. C. Alleman and A. Leeson, eds. Paper from the 4th Int. *In situ* and On-Site Bioremediation Symp. Batelle Press, New Orleans, LA.

Alleman BC and Leeson A (1999). *In situ* bioremediation of petroleum hydrocarbons and other organic compounds. Battelle Press, San Diego.

Anonymous (2001) Water, Environment, Landscape Management at Contaminated Megasites (WELCOME). Description of Work. 5th Framework Programme of the European Union. EESD/RTD Programme, Contract nr. EVK1-2001-00132.

Beeman RE, Howell JE, Shoemaker SH, Salazar EA and Buttram JR (1994) A field evaluation of *in situ* microbial reductive dehalogenation by the transformation of chlorinated ethenes. Pages 14-27 in R. E. Hinchee, A. Leeson, L. Semprini, and S. K. Ong, (eds) Bioremediation of Chlorinated and Polycyclic Aromatic Hydrocarbon Compounds. Lewis, Boca Raton, FL.

de Bruin WP, Kotterman MJJ, Posthumus MA, Schraa G and Zehnder AJB (1992) Complete biological reductive transformation of tetrachloroethene to ethane. Appl. Environ. Microbiol. 58: 1996-2000.

Hendrickson, E.R. S, M.G., Elberson MA, Payne JA, Mack EE, Huang H-B, McMaster ML and Ellis DE (2001). Using a molecular approach to monitor a bioaugmentation. Battelle Press, Columbus Richland.

Langenhoff AAM, Van Liere HC, Pijls CGJM, Schraa G and Rijnaarts HHM (1999) Combined intrinsic and stimulated *in situ* biodegradation of hexachlorocyclohexane (HCH). Pages 81-87 in A. Leeson and B. C. Alleman, (eds) Phytoremediation and innovative strategies for specialised remedial applications, The fifth international *in situ* and on site bioremediation symposium. Battelle Press, San Diego.

Langenhoff AAM, Nipshagen AAM, Bakker C, Krooneman J and Visscher G (2001a) Monitoring stimulated reductive dechlorination at the Rademarkt in Groningen, The Netherlands. Pages 141-147 in A. Leeson and B. C. Alleman, (eds) Anaerobic degradation of chlorinated solvents. Battelle Press, San Diego.

Langenhoff AAM, Staps S, Rijnaarts H, Praamstra T, Heijnen M and Vis P (2001b) Feasibility study of a "Biological Fence" at the site of Shell Netherlands Refinery. TNO R2001/314, Apeldoorn, The Netherlands.

Leeson A, Johnson PC, Hinchee RE, Semprini L and Magar VS (2001). *In situ* aeration and aerobic remediation. Battelle Press, San Diego.

McCarty PL and Wilson JT (1992) Natural anaerobic treatment of a TCE plume, St. Joseph, Michigan, NPL site. Pages 47-50 Bioremediation of Hazardous Wastes. USEPA Center for Environmental Research Information, Cincinnatti, OH.

Middeldorp PJM, Jaspers M, Zehnder AJB and Schraa G (1996) Biotransformation of alpha-, beta-, gamma-, and delta-hexachlorocyclohexane under methanogenic conditions. Environ. Sci. Technol. 30: 2345-2349.

Middeldorp PJM, Luijten MLGC, van de Pas B, van Eekert MHA, Kengen SWM, Schraa G and Stams AJM (1999) Anaerobic microbial reductive dehalogenation of chlorinated ethenes. Biorem. J. 3: 151-170.

Morasch B, Richnow HH, Schink B and Meckenstock RU (2000) *In situ* quantification of aerobic and anaerobic toluene degradation by isotope fractionation, poster VAAM 2000.

Nipshagen AAM, Krooneman J, Tuinstra A and Langenhoff AAM (1999) Degradation of per- and trichloroethene under sequential redox conditions - Phase 2: Additional column experiments and aerobic field trial. CUR/SKB 95-1-41, Gouda, The Netherlands.

Norris RD (2001). Technologies competitive with enhanced bioremediation of source zones. Battelle Press, Columbus Richland.

Reineke W and Knackmuss H-J (1984) Microbial metabolism of haloaromatics: Isolation and properties of a chlorobenzene-degrading bacterium. Appl. Environ. Microbiol. 47: 395-402.

Rittmann BE (2000). Natural Attenuation for Groundwater Remediation. National Academy Press, Washington DC.

Semprini L, Hopkins GD, Roberts PV and McCarty PL (1992) *In situ* transformation or carbon tetrachloride and other halogenated compounds resulting from biostimulation under anoxic conditions. Environ. Sci. Technol. 26: 2454-2461.

Smittenberg J, Gerritse J, Van Bemmel JBM, Van den Brink C and Hageman F (2000) Combined remediation phase 2, Implementation phase. CUR/NOBIS 98-1-24, Gouda, The Netherlands.

Spuij F, Alphenaar A, de Wit H, Lubbers R, van de Brink K, Gerritse J, Gottschal J and Houtman S (1997) Full-scale application of *in situ* bioremediation of PCE-contaminated soil. Pages 431-437 in B. C. Alleman and A. Leeson, eds. Paper from the 4th Int. *In situ* and On-Site Bioremediation Symp. Battelle Press, New Orleans, LA.

Suarez MP and Rifai HS (1999) Biodegradation rates for fuel oxygenates and chlorinated solvents in groundwater. Biorem 3: 337-362.

USEPA (2001) Abstracts of Remediation Case Studies, Volume 5. EPA 542-R-01-008.

Van Aalst-Van Leeuwen MA, Brinkman J, Keuning S, Nipshagen AAM and Rijnaarts HHM (1997) Degradation of per- and trichloroethene under sequential redox conditions - Phase 1: Deliverable 2.6; Field characterisation and laboratory experiments. CUR/NOBIS 95-1-41, Gouda.

Van Eekert MHA, Van Ras NJP, Mentink GH, Rijnaarts HHM, Stams AJM, Field JA and Schraa G (1998). Anaerobic biotransformation of beta-hexachlorocyclohexane by methanogenic granular sludge and soil microflora. Environ. Sci. Technol. 32: 3299-3304.

Van Liere H, Gerritse J and Rijnaarts HHM (1999) Diagnosis of combined natural biological degradation of PCE/TCE and BTEX. TNO-MEP R 99/028, Apeldoorn.

Wiedemeier TH, Wilson JT and Kampbell DH (1996) Natural attenuation of chlorinated aliphatic hydrocarbons at Plattsburgh Air Force Base, New York. Pages 35-59 Symposium on Natural Attenuation of Chlorinated Organics in Ground Water, Dallas, TX. USEPA, Office of Research and Development EPA/540/R96/509, Washington DC.

PART 4
IMMOBILISATION OF POLLUTANTS
IN THE SOIL

IMMOBILISATION OF PESTICIDES IN SOIL THROUGH ENZYMATIC REACTIONS

JEAN-MARC BOLLAG
*Laboratory of Soil Biochemistry, Center for Bioremediation and
Detoxification, The Pennsylvania State University, University Park,
Pennsylvania, U.S.A.*

Abstract

Immobilisation phenomena occurring in soil are of great environmental importance because they may lead to a considerable reduction in the bioavailability of pesticides. Both enzymes and abiotic catalysts can mediate the immobilisation process. One of the most important catalytic reactions in soil is oxidative coupling that links naturally occurring and xenobiotic chemicals, such as pesticides, to organic matter. The reaction may be caused by oxidoreductases and may have a detoxification effect. Therefore, pesticide immobilisation through binding to soil constituents can be considered an alternative method of pollution control.

1. Introduction

Immobilisation processes observed in soil are of great environmental significance as they may lead to a considerable reduction in the bioavailability and degradation of pesticides (Alexander 1995; Hatzinger and Alexander 1995). The ability of soils to retain pesticides is attributed to adsorption phenomena and chemical reactions occurring on active surfaces of humus and mineral particles; pesticides can also be retained through entrapment within the soil matrix (Calderbank 1989; Dec and Bollag 1997). According to the proposed models of biodegradation, pesticide molecules must be present in the aqueous phase to be available to microorganisms. This bioavailability requirement is constantly challenged in terrestrial systems where pesticide molecules are continuously removed from the soil solution through immobilisation or diffusion into inaccessible locations.

Bound residues, by definition, cannot be removed from soil by non-destructive methods (Roberts *et al.* 1984). Pesticides retained through adsorption are not considered bound as they can be desorbed by extraction with water or organic solvents. Pesticides are, in general, adsorbed faster than they are desorbed, a phenomenon known as hysteresis. According to recent observations, the rates of adsorption and desorption are

S.N. Agathos and W. Reineke (eds.).
Biotechnology for the Environment: Soil Remediation, 93-101.
© 2003 Kluwer Academic Publishers. *Printed in the Netherlands.*

subject to reduction with the length of time that the xenobiotics reside in soil or are "aged" (Pignatello and Xing 1996). As they age, chemicals show increased hysteresis. In addition, considerable amounts of aged xenobiotics become entirely resistant to desorption and thus are unable to be biodegraded (Hatzinger and Alexander 1995).

Ageing is currently ascribed to sorption and entrapment phenomena occurring in remote micro-sites within the soil matrix. According to Alexander (1995), xenobiotics can reach these sites by diffusion across the organic matter. Such a combination of diffusion processes with adsorption frequently is referred to as *sequestration* or *slow sorption*, because months or even years may be required to reach equilibrium. Like adsorbed chemicals, those that are sequestered can be recovered, although with difficulty, by vigorous extraction with organic solvents. From the standpoint of biodegradation, however, sequestration is practically irreversible, because the chemicals involved in diffusion and subsequent sorption do not desorb back into the soil solution.

It is well known that a large portion (20 to 70%) of a particular chemical that reaches the terrestrial system becomes sequestered or bound to soil and resists extraction with water and organic solvents (Calderbank 1989). Neither research in the laboratory nor practical experience has found any significant any negative or toxic impact of bound xenobiotics on the environment; consequently the immobilisation of pollutants in soil has been recognised as a promising decontamination technique.

2. Reactions between pesticides and humic material

Binding to humus constitutes one of the major reactions by which anthropogenic compounds are transformed in nature. As outlined in Section 1, pesticides interact with soil colloids through several mechanisms; a number of reviews describe the various mechanisms (Pignatello 1989; Koskinen and Harper 1990). Adsorption occurs primarily as a consequence of the attraction between the solid surface of the soil and the soluble or vapour phase of the pesticide. The nature and strength of adsorption depend largely on the chemical structure of the molecule. Adsorption is reversible, and the desorbed chemicals are available to interact with the biota. However, there is abundant evidence suggesting that with longer exposure to soil, adsorbed residues become more resistant to extraction and degradation (Calderbank 1989; Hatzinger and Alexander 1995). This resistance may result from a gradual sequestration or slow incorporation of the pollutant into humus.

2.1. COVALENT BINDING BY SOIL MICROORGANISMS

The most persistent complexes result from the covalent binding of xenobiotics to humic material. These complexes, often referred to as the "bound residues", are highly resistant to acid and base hydrolysis, thermal treatment, and microbial degradation (Helling and Krivonak 1978; Katan and Lichtenstein 1977; Roberts 1984). In a sense, bound residues constitute a dead-end product of microbial activity. Microorganisms and their enzymes may, in fact, be indispensable in bound residue formation. Experiments using ^{14}C-labelled compounds demonstrated that generally only negligible amounts of bound residues are formed in sterile soils. The role of microorganisms in these processes is to condition the xenobiotic molecules for covalent binding.

Microorganisms also can partially degrade xenobiotics, thus converting them to more reactive derivatives that may be involved in future covalent binding. As a result of binding, however, these derivatives are resistant to further degradation or mineralisation on exposure to microbial populations.

2.2. OXIDATIVE COUPLING

Oxidative coupling is one of the most important chemical reactions occurring in soil (Bollag *et al.* 1997). It leads to humification of both naturally occurring humic acid precursors and susceptible anthropogenic compounds (phenols and aromatic amines) through their incorporation into soil organic matter. The incorporation is controlled by a free radical mechanism. The resonance-stabilised free radicals, formed through the loss of an electron and a proton from a phenol molecule, couple to each other in a variety of combinations. After coupling, phenolic moieties are mostly linked through C-C and C-O bonds, whereas aromatic amines form C-N and N-N linkages (Sjoblad and Bollag 1981).

Oxidative coupling is mediated by a number of biological and abiotic catalysts, including microbial or plant enzymes, inorganic chemicals (e.g., ferric chloride, cupric hydroxide) and clay minerals (Bollag 1983; Wang *et al.* 1986). Coupling reactions can also occur spontaneously in the presence of oxygen at neutral and alkaline pH values (Musso 1967). Spontaneous reactions frequently lead to the incorporation of non-phenolic compounds into humic polymers.

3. Enzymes and their origin

Many soil microorganisms produce extracellular oxidoreductases capable of catalysing the coupling of aromatic compounds. These enzymes are classified as either peroxidases or polyphenol oxidases.

3.1. PEROXIDASES

All peroxidases contain an iron porphyrin ring and require the presence of peroxides (e.g., hydrogen peroxide) for activity. In particular horseradish peroxidase (HRP), which catalyses the polymerisation of a wide range of phenolic and aniline compounds, was tested for the detoxification of industrial wastewater (Klibanov *et al.* 1980; Maloney *et al.* 1986; Dec and Bollag 1994).

3.2. POLYPHENOL OXIDASES

The polyphenol oxidases are divided into two groups: laccases and tyrosinases, which require bimolecular oxygen, but no coenzyme, for activity. However, the enzymes differ in the mechanism by which they oxidise phenols. Laccases oxidise phenolic compounds to form their corresponding anionic free radicals, whereas tyrosinases form *o*-diphenols and subsequently release oxidised *o*-quinones (Sjoblad and Bollag 1981). In an alkaline environment, the quinone products slowly polymerise through autooxidative processes.

The laccases may prove to be the most useful of the phenoloxidases because, like the peroxidases, they produce highly reactive radicals, but unlike the latter, they do not require the presence of hydrogen peroxide.

3.3. FUNCTION OF ENZYMES IN BINDING REACTIONS BETWEEN PESTICIDES AND HUMIC MATERIAL

The main role of enzymes or minerals in oxidative coupling is to mediate the oxidation of the substrates to free radicals (Musso 1967; Bollag 1992). Once the free radicals are generated, coupling is completed without further involvement of the catalyst. In modelling studies we investigated the catalytic effects of enzymes on pesticides or their phenolic or anionic intermediates by determining their binding to humic material, and whenever possible, we identified the resulting oxidative coupling products.

On the basis of early observations, xenobiotic anilines and phenols were assumed to form covalent linkages with soil organic matter (Hsu and Bartha 1976; Bollag *et al.* 1980; Parris 1980). It was well understood that the validity of these assumptions could be verified only through direct inspection of the intact complexes. However, the intricate and heterogeneous structure of humic substances made it difficult to achieve direct insight. To overcome this problem, complex soil systems were replaced by simple models in which the xenobiotic chemicals were allowed to interact with the monomeric constituents of humic acid in the presence of enzymes. Using model substrates, the resulting products were relatively easy to isolate from the reaction mixture and could be analysed for their exact molecular configuration.

In a later study, Simmons *et al.* (1989) observed only partial oxidation of 4-chloroaniline in the presence of various catalysts. The resulting free radicals first bound to each other to form a dimer that subsequently was condensed with a resonance-stabilised guaiacol anion. The aniline molecules that did not undergo oxidation were subject to Michael addition to quinone oligomers resulting from the coupling of the guaiacol free radicals. In the study of Tatsumi *et al.* (1994), free radical formation was limited exclusively to ferulic acid, and all aniline molecules were incorporated into the resulting dimers through a non-radical condensation.

In contrast to anilines, xenobiotic phenols first had to be enzymatically oxidised to aryloxy-free radicals or quinones to become bound to humic substances. For example, when Sarkar *et al.* (1988) incubated 2,4-dichlorophenol with fulvic acid, no binding occurred in the absence of catalysts; however, considerable binding was observed upon the addition of oxidoreductases such as mushroom tyrosinase, horseradish peroxidase, or the laccases from *Trametes versicolor* or *Rhizoctonia praticola*.

4. NMR spectroscopy to determine the type of binding of pesticides in the soil

Historically, research on pesticide immobilisation was carried out using ^{14}C-labelled chemicals combined with radiation counting. Recently, labelling of pollutants with stable isotopes (^{13}C or ^{15}N) combined with nuclear magnetic resonance (NMR) spectroscopy emerged as a non-destructive technique with great identification potential. The major advantage of this approach lies in the fact that pollutant molecules enriched with the ^{13}C or ^{15}N isotope generate more intensive NMR signals than those resulting from the natural abundance of ^{13}C or ^{15}N in the studied compound and soil. The basis of the NMR approach is that any binding-related modification in the original arrangement of the labelled atoms automatically induces changes in the position of the corresponding signals in the NMR spectra. The delocalisation of the signals exhibits a high degree of specificity, indicating whether or not binding has occurred and, if so, the type of bond formed.

NMR was used successfully to determine covalent binding of pesticides and other xenobiotics (Haider et al. 1993; Wais et al. 1994; Thorn et al. 1996). In the study of Haider et al. (1993), for instance, the fungicide anilazine [4,6-dichloro-N-(2-chloro-phenyl)-2,3,5-triazine-2-amine] labelled with ^{14}C and ^{13}C in the triazine ring was found to be immobilised in soil through ligand exchange. The NMR spectra of humic acid extracted from soil together with the bound fungicide revealed that the chlorine substituents located at the C-4 and C-6 positions were removed from the triazine ring and replaced by the oxygen-containing functional groups of soil organic matter (Wais et al. 1994). This exchange resulted in the formation of strong ether and ester linkages between the dehalogenated anilazine molecule and the humic matrix.

Thorn et al. (1996) used ^{15}N-NMR spectroscopy to demonstrate covalent binding of ^{15}N-labelled aniline to humic acid when the two components were dissolved in water and stirred for 5 days at pH 6. The changes in the chemical shifts of the ^{15}N atom indicated that binding was due to nucleophilic addition reactions of aniline with the quinone or carbonyl groups typical for humic substances. The labelled chemical was incorporated in the form of anilinohydroquinone, anilinoquinone, anilide, imine, and heterocyclic nitrogen. The latter comprised more than 50% of the bound amines.

Our ^{13}C-NMR studies were carried out using two ^{13}C-labelled pollutants: 2,4-dichlorophenol, a degradation product of the herbicide 2,4-D (Hatcher et al. 1993), and cyprodinil, a new phenyl-pyrimidine amine fungicide manufactured by Ciba-Geigy (Dec et al. 1997). 2,4-Dichlorophenol, labelled either in the C-1 or in the C-2 and C-6 position, was incubated for 2 hours with dissolved humic acid in the presence of a peroxidase. The NMR signals generated by the ^{13}C label demonstrated bonding between the two components through carbon-carbon, ester and phenolic ether linkages.

To investigate the formation of covalent bonds under more natural conditions, cyprodinil, which was labelled either in the phenyl or pyrimidyl ring, was incubated with fresh soil for several months (Dec et al. 1997). After exhaustive washing with methanol, humic acid was isolated by extraction with 0.5 M NaOH. Humic acid was then purified by dialysis and/or silylated by treatment with trimethylchlorosilane to facilitate the ^{13}C-NMR analysis. The NMR signals generated by both the dialysed and silylated samples indicated cleavage of the cyprodinil molecule between the aromatic rings and covalent binding of the phenyl and pyrimidyl moieties to humic acid.

Nanny et al. (1997) demonstrated the use of ^{13}C-NMR spectroscopy for the assessment of non-covalent complexes of xenobiotics with humic materials. The compound under investigation (^{13}C-labelled acenaphthenone) and humic material (fulvic acid) were allowed to interact directly in the NMR tube in a methanol/D$_2$O solution. Based on the measurements of the decreasing spin-lattice relaxation time (T_1) of the labelled atom, the authors determined that the sorption of acenaphthenone to fulvic acid involved weak hydrophobic or hydrogen bonding.

The results, especially those related to binding in the soil, confirm the great potential of using ^{13}C-labelled chemicals with ^{13}C-NMR spectrometry as an analytical technique for evaluating interactions between pollutants and soil components, particularly humic substances. Establishing the physical or chemical associations between pollutants and humic substances will provide essential data for understanding principles of bioavailability.

5. Stability and release of bound pesticides

Before the binding of xenobiotics to humus can be applied as a decontamination procedure, the stability of the bound complexes must be investigated. If large quantities of a bound pollutant were released at a future time, the accumulation of these complexes would pose a delayed environmental hazard. The stability of humus-bound xenobiotics has been demonstrated by several investigators. In one study, when 3,4-dichloroaniline was applied to a German soil, 46% of the compound remained bound to the soil 2 years after treatment (Viswanathan *et al.* 1978). In a separate study, 83% of ^{14}C-labelled-atrazine remained associated with the soil after 9 years; 50% of this residue represented bound material (Capriel *et al.* 1985).

The activity of microorganisms is believed to be the primary factor responsible for the release of bound residues. To study the release of bound pesticides, ^{14}C-labelled-catechol and mono-, di-, tri-, and pentachlorophenols, bound to humic acid polymers, were incubated with microbial soil populations, and the release of radioactive compounds into the medium was monitored for 13 weeks (Dec and Bollag 1988). This study demonstrated that ^{14}C-labelled compounds were released in small quantities and the release was accompanied by a simultaneous mineralisation of the bound material to ^{14}CO$_2$. As might be expected, the release of bound xenobiotics differs with the type of binding. A "surface" fraction of bound residue appears to be releasable, whereas the remainder is covalently bound to a "core" portion that is less accessible to microbial degradation.

Overall, the available data indicate that the microbial release of bound xenobiotics occurs at an extremely slow rate. Once released, the xenobiotics can be mineralised or reincorporated into humus. Consequently, released residues are not expected to accumulate and should not pose a delayed health hazard.

There is little doubt that binding interactions reduce bioavailability and considerably contribute to the recalcitrance of xenobiotics in soil. However, taking into account that a given chemical may undergo binding by several mechanisms simultaneously and that xenobiotic molecules frequently compete for binding sites with other chemicals present in soil environments, it will never be easy to determine the exact effect of bioavailability on xenobiotic recalcitrance in *in vivo* situations.

6. Enzymes as decontaminating agents

The use of microbial extracellular enzymes as decontaminating agents is hindered by the fact that free enzymes are rapidly inactivated under the harsh conditions of soil. This problem may be overcome by immobilising the free enzymes on solid supports.

Immobilisation has been shown to enhance the thermostability of the enzymes, to prevent their degradation by proteases, and to increase their half-life (Sarkar *et al.* 1989). However, the enzymatic activity retained after immobilisation depends on the method of immobilisation and the nature of the solid support.

The activity of laccase immobilised on soil supports was investigated using radiolabelled 2,4-DCP (Ruggiero *et al.* 1989). The efficiency of laccase-kaolinite and laccase-soil complexes in removing 2,4-DCP was equivalent to that of the free enzyme (~95%). Laccases bound to montmorillonites 1 and 2 were less efficient, with removal activities of 69% and 42%, respectively. However, after 6 hours of incubation, the laccases, immobilised on soil particles, retained 86 to 100% of their activity, whereas the free enzyme retained only 62% of its activity. Furthermore, the immobilised enzyme could be recovered from the reaction mixture and used repeatedly to transform the substrate, with minimum loss of activity. Overall, because of their increased biochemical stability and reusability, immobilised enzymes are more cost effective than are free enzymes.

7. Conclusions

The incorporation of pesticides and their derivatives into soil organic matter occurs readily in nature. This process can be used to immobilise and detoxify hazardous compounds. The immobilisation phenomena have several important consequences: (1) the amount of compound available to interact with the biota is reduced; (2) the complexed products are less toxic than their parent compounds; and (3) binding restricts leaching of chemicals across the soil profile, thus preventing groundwater contamination.

The use of enzymatic coupling for detoxification raised some concerns about the ultimate fate of bound pesticides. However, all available data indicate that, once xenobiotics are incorporated into soil, they are released at a very slow rate and to a minimal extent. The gradual release should not pose a delayed health hazard because the released compounds can be mineralised to CO_2 or reimmobilised in the soil matrix.

Prior to the practical application of enzymatic treatment, extensive research is required to further analyse the accumulation, bioavailability, and toxicity of bound pesticide residues under field conditions. Research should continue to focus on the development of new methods for maximising the binding process, e.g., through the use of immobilised enzymes, abiotic catalysts, or co-polymerising agents. All current evidence indicates that the enzymatic or abiotic incorporation of xenobiotics into humus provides an efficient method for detoxifying hazardous pollutants.

References

Alexander, M. (1995) How toxic are toxic chemicals in soil?, Environ. Sci. Technol. 29, 2713-2717.

Bollag, J.-M. (1983) Cross-coupling of humus constituents and xenobiotic substances, in R. Christman and E. Gjessing (eds.), Aquatic and Terrestrial Humic Substances, Ann Arbor Science Publishers, Inc., pp. 127-141.

Bollag, J.-M. (1992) Decontaminating soil with enzymes, Environ. Sci. Technol. 26, 1876-1881.

Bollag, J.-M., Dec, J., and Huang, P.M. (1997) Formation mechanisms of complex organic structures in soil habitats, Adv. Agron. 63, 237-266.

Bollag, J.-M., Liu, S.-Y., and Minard, R. (1980) Cross-coupling of phenolic humus constituents and 2,4-dichlorophenol, Soil Sci. Soc. Amer. J., 44, 52-56.

Calderbank, A. (1989) The occurrence and significance of bound pesticide residues in soil, Rev. Environ. Contam. Toxicol. 108, 71-103.

Capriel, P., Haisch, A., and Kahn, S. (1985) Distribution and nature of bound (non-extractable) residues of atrazine in a mineral soil nine years after herbicide application, J. Agric. Food Chem. 33, 567-569.

Dec, J. and Bollag, J.-M. (1988) Microbial release and degradation of catechol and chlorophenols bound to synthetic humic acid, Soil Sci. Soc. Amer. J. 52, 1366-1371.

Dec, J. and Bollag, J.-M. (1994) Use of plant material for the decontamination of water polluted with phenols, Biotechnol. Bioeng. 44, 1132-1139.

Dec, J. and Bollag, J.-M. (1997) Determination of covalent and non-covalent binding interactions between xenobiotic chemicals and soil, Soil Sci. 162, 858-874.

Dec, J., Haider, K., Benesi, A., Rangaswamy, V., Schäffer, A., Plücken, U., and Bollag, J.-M. (1997) Analysis of soil-bound residues of the ^{13}C-labeled fungicide cyprodinil by NMR spectroscopy, Environ. Sci. Technol. 31, 1128-1135.

Haider, K., Spiteller, M., Wais, A. and, Fild, M. (1993) Evaluation of the binding mechanism of anilazine and its metabolites in soil organic matter, Intern. J. Environ. Anal. Chem. 53, 125-137.

Hatcher, P., Bortiatynski, J., Minard, R., Dec, J., and Bollag, J.-M. (1993) Use of high-resolution ^{13}C NMR to examine the enzymatic covalent binding of ^{13}C-labeled 2,4-dichlorophenol to humic substances, Environ. Sci. Technol. 27, 2098-2103.

Hatzinger, P. and Alexander, M. (1995) Effect of aging of chemicals in soil on their biodegradability and extractability, Environ. Sci. Technol. 29, 537-545.

Helling, C. and Krivonak, A. (1978) Physiochemical characteristics of bound dinitroaniline herbicides in soil, J. Agric. Food Chem. 26, 1156-1163.

Hsu, T.-S. and Bartha, R. (1976) Hydrolysable and non-hydrolysable 3,4-dichloroaniline humus complexes and their respective rates of biodegradation, J. Agric. Food Chem. 24, 118-122.

Katan, J. and Lichtenstein, E.P. (1977) Mechanisms of production of soil-bound residues of [14C] parathion by microorganisms, J. Agric. Food Chem. 25, 1404-1408.

Klibanov, A., Alberti, B., Morris, E., and Felshin, L. (1980) Enzymatic removal of toxic phenols and anilines from wastewaters, J. Appl. Biochem. 2, 414-421.

Koskinen, W.C. and Harper, S.S. (1990) The retention process: Mechanisms, in H.H. Cheng (ed.), Pesticides in the Soil Environment: Processes, Impacts, and Modeling, Soil Sci. Soc. of Am., Inc., Madison, Wisconsin, pp. 51-77.

Maloney, S., Manem, J., Mallevialle, J., and Fiessinger, F. (1986) Transformation of trace organic compounds in drinking water by enzymatic oxidative coupling, Environ. Sci. Technol. 20, 249-253.

Musso, H. (1967) In W. Taylor and A. Battersby (eds.), Oxidative Coupling of Phenols, Marcel Dekker, New York.

Nanny, M.A., Bortiatynski, J.M., and Hatcher, P.G. (1997) Non-covalent interactions between acenaphthenone and dissolved fulvic acid as determined by ^{13}C NMR T_1 relaxation measurements, Environ. Sci. Technol. 31, 530-534.

Parris, G.E. (1980) Covalent binding of aromatic amines to humates. 1. Reactions with carbonyls and quinones, Environ. Sci. Technol. 14, 1099-1106.

Pignatello, J. (1989) Reactions and movement of organic chemicals in soils, in B. Sawhney and K. Brown (eds.), Sorption Dynamics of Organic Compounds in Soils and Sediments, Spec. Publ. No. 22, Soil Sci. Soc. Am., Inc., Madison, Wisconsin, pp. 45-80.

Pignatello, J.J. and Xing, B. (1996) Mechanisms of slow sorption of organic chemicals to natural particles, Environ. Sci. Technol. 30, 1-11.

Roberts, T., Klein, W., Still, G.G., Kearney, P.C., Drescher, N., Desmoras, J., Esser, H.O., Aaharonson, N., and Vonk, J. (1984) Non-extractable pesticide residues in soils and plants, Pure Appl. Chem. 56, 945-956.

Ruggiero, P., Sarkar J. M., and Bollag, J.-M. (1989) Detoxification of 2,4-dichlorophenol by a laccase immobilized on soil or clay, Soil Sci. 147, 361-370.

Sarkar, J., Malcolm, R., and Bollag, J.-M. (1988) Enzymatic coupling of 2,4-dichlorophenol to stream fulvic acid in the presence of oxidoreductases, Soil Sci. Soc. Am. J. 52, 688-694.

Sarkar, J. M., Leonowicz, A., and Bollag, J.-M. (1989) Immobilization of enzymes on clays and soils, Soil Biol. Biochem. 21, 223-230.

Simmons, K.E., Minard, R.D., and Bollag, J.-M. (1989) Oxidative co-oligomerisation of guaiacol and 4-chloroaniline, Environ. Sci. Technol. 23, 115-121.

Sjoblad, R. and Bollag, J.-M. (1981) Oxidative coupling of aromatic compounds by enzymes from soil microorganisms, in E. Paul and J. Ladd (eds.), Soil Biochemistry, vol. 5, Marcel Dekker Inc., New York, pp. 113-152.

Tatsumi, K., Freyer, A., Minard, R., and Bollag, J.-M. (1994) Enzyme-mediated coupling of 3,4-dichloroaniline and ferulic acid: A model for pollutant binding to humic materials, Environ. Sci. Technol. 28, 210-215.

Thorn, K.A., Pettigrew, P.J., Goldenberg, W.S. and Weber, E.J. (1996) Covalent binding of aniline to humic substances. 2. ^{15}N NMR studies of nucleophilic addition reactions, Environ. Sci. Technol. 30, 2764-2775.

Viswanathan, R., Scheunert, I., Kohli, J., Klein, W. and Korte, F. (1978) Long- term studies on the fate of 3,4-dichloroaniline-^{14}C in a plant-soil system under outdoor conditions, J. Environ. Sci. Health B, 13, 243-259.

Wais, A., Burauel, E., de Graaf, A.A., Haider, K., and Führ, F. (1994) ^{13}C-NMR- und GPC-Untersuchungen an Modellhuminsäuren aus Maisstroh als Basis zur Spezifizierung von Bindungsformen nichtextrahierbarer Rückstände von Xenobiotika in Böden, Mitt. Deutsch. Bodenkdl. Ges. 74, 485-488.

Wang, T., Huang, P., Chou, C.-H., and Chen, J.-H. (1986) Interactions of Soil Minerals with Natural Organics and Microbes, Spec. Publ. No. 17, Soil Sci. Soc. Am., Inc., Madison, Wisconsin, pp. 251-281

HUMIFICATION OF NITROAROMATICS

DIRK BRUNS-NAGEL, HEIKE KNICKER, OLIVER DRZYZGA,
EBERHARD VON LÖW, KLAUS STEINBACH, DIETHARD
GEMSA
*DB-N, EvL, KS, DG: Institute of Immunology and Environmental
Hygiene, University of Marburg, Pilgrimstein 2, D-35037 Marburg,
Germany, Fax:+49/06421/2862309, umwhy@post.med.uni-marburg.de,
current address: Aventis Behring GmbH, Emil-von-Behring-Straße 76, D-
35041 Marburg, Germany, Dirk.Bruns-Nagel@aventis.com; HK:
Department of Soil Science, TU München, D-85350 Freising-
Weihenstephan, Germany; OD: University of Groningen, Kerklaan 30,
NL-9751 NN Haren, The Netherlands*

1. Introduction

Today nitroaromatic compounds and metabolites are ubiquitous contaminants of the environment. Dinitrotoluenes, e. g. are produced in huge amounts for the manufacture of polyurethane foams, drugs, and dyes. Furthermore, different pesticides and fragrances are nitroaromatics and pollute the environment. Last, but not least, some major explosives are nitroaromatic compounds.

Worldwide, 2,4,6-trinitrotoluene (TNT) and related compounds contaminate soil and groundwater of former ammunition plants. The toxicity of the nitroaromatics (Tan *et al.* 1992, Drzyzga *et al.* 1995, Honeycutt *et al.* 1996, Neumann 1996, Bruns-Nagel *et al.* 1999a) and the fact that polar, highly water soluble metabolites can be generated in soil (Bruns-Nagel *et al.* 1999b) and threaten the groundwater indicate that a cleanup of contaminated areas is urgently needed. We developed an anaerobic/aerobic composting system for bioremediation of TNT-contaminated soil. The two-step system leads to a humification of the explosive. New results concerning the transformation of TNT under composting conditions, which lead to a better understanding of the humification process, will be presented.

2. Composting of soil contaminated with nitroaromatics

Composting of contaminated soil represents an effective low cost technique of bioremediation. The general strategy is to mix contaminated soil with a substrate that

S.N. Agathos and W. Reineke (eds.).
Biotechnology for the Environment: Soil Remediation, 103-111.
© 2002 Kluwer Academic Publishers. Printed in the Netherlands.

can be composted. During the composting of the substrate, the soil contamination is mineralised or transformed by the soil microorganisms.

Different researchers used composting as bioremediation technique for TNT-contaminated soil (Kaplan and Kaplan 1982, Isbister *et al.* 1984, Griest *et al.* 1993). They all found that the process significantly reduced the extractable amount of TNT and related compounds. Furthermore, previous experiments with ^{14}C-TNT revealed that the explosive is very recalcitrant towards a microbial mineralisation.

In our research we improved the composting process, investigated the TNT transformation pathway during the composting process, and explored the character of non-extractable explosive residues formed during composting.

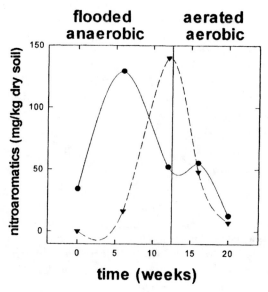

time (weeks)

● **sum 4- and 2-ADNT** ▼ **sum 2,4- and 2,6-DANT**

Figure 1: Time course of aromatic amines formed during an anaerobic/aerobic composting process of TNT-contaminated soil. As substrate, sugar beet and straw was used. 4- and 2-ADNT, 4- and 2-aminodinitrotoluene; 2,4- and 2,6-DANT, 2,4- and 2,6-diaminonitrotoluene.

3. Optimisation of composting of TNT-contaminated soil

In our laboratory, different substrates were tested for soil composting. Chopped sugar beet turned out to be a highly efficient additive to compost soil. The aerated incubation of contaminated soil amended with 80% (v/v) sugar beet caused a decline of the extractable amount of TNT and metabolites from about 7 000 mg/kg dry compost to about 400 mg/kg dry compost within only 28 days (Breitung *et al.* 1996). However, a drastic treatment of the composted soil with 8 M HCl revealed that this destructive chemical caused the mobilisation of up to 3 000-mg nitroaromatics/kg dry compost. This indicated that the classical aerobic composting process was not suitable for bioremediation of TNT-contaminated soil.

The composting procedure was decisively improved by performing an anaerobic phase before starting the aerobic composting process. Experiments were performed with different soil/sugar beet and soil/sugar beet/straw mixtures (Breitung *et al.* 1996, Bruns-Nagel *et al.* 1997). To generate anaerobic conditions, the compost mixtures were flooded or water-saturated for several weeks prior to the aerated composting phase. This process caused a reduction of the extractable amount of nitroaromatics of 98-99% within 20 weeks (Bruns-Nagel *et al.* 1997). A characteristic time course for the generation and disappearance of TNT transformation products was seen during this treatment (Figure 1).

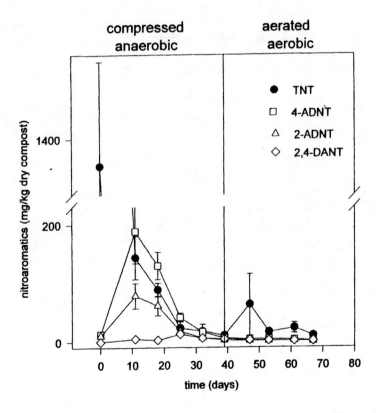

Figure 2: Anaerobic/aerobic composting of TNT-contaminated soil in a laboratory scale experiment. Each data point represents the mean of five parallel compost extractions. Bars indicate the standard deviation. TNT, 2,4,6-trinitrotoluene; 4-ADNT, 4-amino-2,6-dinitrotoluene; 2-ADNT, 2-amino-4,6-dinitrotoluene; 2,4-DANT, 2,4-diamino-6-nitrotoluene.

Even a drastic extraction of the anaerobic/aerobic composted soil with HCl did not lead to a mobilisation of TNT or any transformation products.
The anaerobic/aerobic composting system had two major disadvantages:

• sugar beet is not always available during the year,

• an expensive drying process might be necessary after the flooded phase.

These disadvantages could be overcome by an improved composting procedure, which was developed by our institute and the company Plambeck ContraCon. Decisive changes were that a substrate was used which is available at any time of the year and that the flooding was substituted by compressing the compost mixture (Winterberg *et al.* 1998). The improved anaerobic/aerobic composting system yielded a reduction of the extractable amount of TNT by 99% and more. Figure 2 shows the result of a laboratory experiment in a bioreactor filled with about 3 kg compost mixture.

A characteristic transformation of TNT to mono- and diaminonitrotoluenes can be observed, followed by a disappearance of these products. In contrast to the results shown in Figure 1 these metabolites already fade away during the anaerobic phase. This was probably due to the fact that the compost material was stirred up once a week which introduced air into the compost.

Further experiments with dinitrotoluene (DNT)-contaminated soil showed similar results. Furthermore, it could be demonstrated, that the process also eliminated the recently identified nitrobenzoic acids in TNT-contaminated soil (Bruns-Nagel *et al.* 1999b).

Our results and the fact that TNT is very recalcitrant against a microbial mineralisation imply that the elimination of the nitroaromatics is based on a transformation (reduction) of the nitroaromatic compounds and a subsequent humification of the aromatic amines. A simplified process scheme of the anaerobic/aerobic composting treatment is given in Figure 3.

4. Transformation of TNT during anaerobic/aerobic composting

Besides the transformation of TNT to well-known transformation products like amino- and diaminonitrotoluenes, we were able to identify three different conjugated products formed during the anaerobic/aerobic composting process (Table 1) (Bruns-Nagel *et al.* 1998a). 4-*N*-acetylamino-2-hydroxylamino-6-nitrotoluene (4-*N*-AcOHANT) and 4-*N*-formylamino-2-amino-6-nitrotoluene (4-*N*-FAmANT) were only generated under anaerobic conditions, whereas 4-*N*-acetylamino-2-amino-6-nitrotoluene (4-*N*-AcANT) was synthesised under both aerobic and anaerobic conditions.

All these metabolites did not accumulate as dead-end products. For 4-*N*-AcANT it could be shown that the acetyl group was easily removed under aerobic conditions which led to the formation of 2,4-diaminonitrotoluene (2,4-DANT). The degradation pathway of the two other metabolites remains unknown.

The significance of the formation of conjugated TNT metabolites is not known at the present time. It might be a detoxification reaction, as postulated for aniline (Tweedy *et al.* 1970). Furthermore, 4-*N*-FAmANT is a known intermediate product of the white rot fungus *Phanaerochaete chrysosporium* during the formation of 2,4-DANT (Michels and Gottschalk 1995).

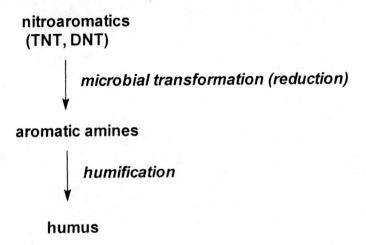

Figure 3: *Scheme of the detoxification of soil contaminated with nitroaromatics via an anaerobic/aerobic composting process.*

Since at present almost nothing is known about the toxicity of these metabolites, it is strongly recommended to analyse bioremediated soils for these metabolites. In addition, research is needed to understand the importance of conjugated metabolites.

5. ^{14}C-TNT balancing in anaerobic/aerobic composting

As shown in Figure 3, it is assumed that the elimination of nitroaromatics via an anaerobic/aerobic composting process is finally caused by a humification of microbially formed aromatic amines. To investigate the mass flow and stability of humified TNT metabolites, studies with ^{14}C-TNT were performed. ^{14}C-TNT spiked soil from a former ammunition plant was composted anaerobic/aerobic. After each treatment phase, a sample was withdrawn and analysed. First, different extractions were applied. Thereafter, the extracted soil was fractionated into humic and fulvic acid and humin by boiling for 2h in a reflux unit with 50 % (w/v) NaOH. The results of this treatment are shown in Table 2.

After the aerobic phase, 56.8% of the radio-labelled TNT was transformed to a non-extractable residue. The anaerobic/aerobic treatment increased the non-extractable quota to 82.6%. Furthermore, the drastic treatment (boiling with concentrated NaOH) indicates that the formed residues were bound in a very stable manner.

Table 1: Presently identified acetylated and formylated TNT metabolites in anaerobic/aerobic composts (Bruns-Nagel et al. 1998a)

substance name	chemical structure
4-*N*-acetylamino-2-hydroxylamino-6-nitrotoluene (4-*N*-AcOHANT)	
4-*N*-formylamino-2-amino-6-nitrotoluene (4-*N*-FAmANT)	
4-*N*-acetylamino-2-amino-6-nitrotoluene (4-*N*-AcANT)	

Table 2: Distribution of radioactivity in ^{14}C-TNT spiked soil after an anaerobic and anaerobic/aerobic treatment (Drzyzga et al. 1998).

Sample	Soil Extract			Humic Acid	Fulvic Acid	Humin	Rest
	Ethyl acetate	Methanol	Water				
	% radioactivity						
Anaerobic Phase	9.5	22.1	8.6	4.5	5.3	47.0	3
Anaerobic/ Aerobic Phase	3.3	4.4	1.7	11.1	9.5	62.0	8

6. Qualitative description of non-extractable ^{15}N-TNT residues formed by an anaerobic/aerobic composting

Our previous investigations strongly indicate that the humification of nitroaromatics leads to a very stable incorporation of metabolites of the original xenobiotics into humic structures. We assumed that the amino functions of the aromatic amines were the reactive groups involved in the binding mechanisms. For a better understanding experiments with ^{15}N labelled TNT were performed. This non-radioactive label allows a qualitative description of formed residues via ^{15}N nuclear magnetic resonance (NMR) analysis. For these experiments, soil was spiked with ^{15}N-TNT. Thereafter, the soil was treated according to the anaerobic/aerobic composting process. The treated soil was extracted with an organic solvent. The extracted soil was then fractionated into humic and fulvic acid and humin by boiling with concentrated NaOH (Bruns-Nagel *et al.* 1998b, Knicker *et al.* 1999). An example of a solid state NMR spectrum of a humic acid is shown in figure 4. The spectrum was obtained by accumulating only about 3,500 single measurements in a time of about 30 min. NMR spectra of similar resolution of soil samples not enriched with ^{15}N can only be obtained after measurement times of two or more days. This indicates the high enrichment of ^{15}N in the sample.

The spectrum shows two major peaks appearing in spectra of all the soil fractions that were analysed (data not shown). A tentative assignment of the chemical shift regions to ^{15}N functional groups is given in Table 3.

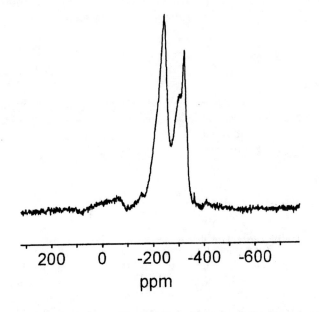

Figure 4: Solid-state ^{15}N NMR spectrum of a humic acid extracted from ^{15}N-TNT spiked soil, which was treated sequentially under anaerobic and aerobic conditions.

Most interestingly, the major peak of the spectrum occurs in the chemical shift region between −165 and −270 ppm. This indicates that more then 50% of the TNT residues are bound in a heterocyclic manner in soil humus compounds. Furthermore, a considerable amount of ^{15}N is present as aniline derivative. The sharp peak at −320 ppm most likely represents free amino groups of TNT reduction products. Since the soil was extracted with solvent prior to the extraction of the humic acid, these amino groups most probably belong to bound TNT metabolites, which contain a further free amino function.

Table 3: Tentative assignment of chemical shift regions to ^{15}N functional groups (Knicker et al. 1999).

Chemical shift region (ppm)	Assignment
148 to 50	azo compounds
50 to −25	nitro groups
-25 to −120	imines, phenoxazinones, pyridines, quinolines
-120 to −165	nitriles, oxazoles
-165 to −270	imidazoles, indoles, pyrroles, carbazoles, quinolone, anilides, enaminones
-270 to −310	aniline derivatives, phenoxazones, hydrazines
-310 to −350	aniline, phenylamines
- 359	ammonium

The formation of heterocyclic compounds and covalent bound aniline derivatives after incubation of ^{15}N aniline with humic compounds was previously reported by Thorn *et al.* (1996). In contrast to these authors, we published for the first time experimental evidence for the formation of these structures under composting conditions (Knicker *et al.* 1999).

7. Conclusion

A highly effective anaerobic/aerobic composting system was developed to detoxify soil contaminated with nitroaromatics. The technique was intensively examined for the remediation of TNT-contaminated soil. It could be shown that, besides the well-known TNT reduction products, further differently conjugated TNT metabolites were generated and degraded under the applied conditions. ^{14}C-TNT investigations revealed that almost 83% of the explosive is bound in a very stable manner to soil humus after an anaerobic/aerobic composting. Additionally performed experiments with ^{15}N-TNT and ^{15}N solid-state NMR analysis strongly indicated that the humification of TNT leads to a covalent and heterocyclic binding of transformation products via the amino functions.

In future research, it seems to be necessary to investigate the long-term stability of the formed humic constituents more closely. This can be carried out by more detailed analyses such as liquid NMR and by the use of reference compounds.

References

Breitung, J., D. Bruns-Nagel, K. Steinbach, L. Kaminski, D. Gemsa, and E.v. Löw (1996) Bioremediation of 2,4,6-trinitrotoluene-contaminated soils by two different aerated compost systems, Appl. Microbiol. Biotechnol. 44, 795-800.

Bruns-Nagel, D., J. Breitung, K. Steinbach, D. Gemsa, E. v. Löw, T. Gorontzy, and K.-H. Blotevogel (1997) Bioremediation of 2,4,6-trinitrotoluene-contaminated soil by anaerobic/aerobic and aerobic methods, in *In situ* and On-Site Bioremediation: Volume 4/2, S. 9-14. Battelle Press, Columbus, Richland, USA.

Bruns-Nagel, D., O. Drzyzga, K. Steinbach, T.C. Schmidt, E. v. Löw, T. Gorontzy, K.-H. Blotevogel, and D. Gemsa (1998a) Anaerobic/aerobic composting of 2,4,6-trinitrotoluene-contaminated soil in a reactor system. Environ. Sci. Technol. 32, 1676-1679.

Bruns-Nagel, D., H. Knicker, O. Drzyzga, B. Casper, K.Steinbach und E. von Löw (1998b) Aufklärung der Humifizierung von TNT mittels stabiler Isotope und NMR-Analytik, Beitrag zum Statusseminar des BMBF-Verbundvorhabens "Langzeit- und Remobilisierungsverhalten von Schadstoffen, 22.-23.September, Bremen, F 1-16, Hrsg.: Umweltbundesamt, Fachgebiet III 3.6

Bruns-Nagel, D., S. Scheffer, B. Casper, H. Garn, O. Drzyzga, E. v. Löw, and D. Gemsa (1999a) Effect of 2,4,6-trinitrotoluene and its metabolites on human monocytes, Environ. Sci. Technol., 33, 2566-2570.

Bruns-Nagel, D.; Schmidt, T.C., Drzyzga, O.; von Löw, E., and K. Steinbach (1999b) Identification of oxidized TNT metabolites in soil samples of a former ammunition plant, Environ. and Pollut. Res., 6, 7-10.

Drzyzga, O., D. Bruns-Nagel, T. Gorontzy, K.-H. Blotevogel, D. Gemsa, and E. v. Löw (1998) Incorporation of ^{14}C-labeled 2,4,6-trinitrotoluene (TNT) metabolites into different soil fractions after anaerobic and anaerobic/aerobic treatment of soil/molasses mixtures, Environ. Sci. Technol. 32 , 3529-3535.

Drzyzga, O., T. Gorontzy, A. Schmidt, and K. H. Blotevogel (1994) Toxicity of explosives and related compounds to the luminescent bacterium *Vibrio fischeri* NRRL-B-11177, Arch. Environ. Contam. Toxicol. 28, 229-235.

Griest, W.H., A.J. Stewart, R.L. Tyndall, J.E. Caton, C.-H. Ho, K.S. Ironside, W.M. Caldwell, and E. Tan (1993) Chemical and toxicological testing of composted explosives-contaminated soil, Environ. Toxicol. Chem.,12, 1105-1116.

Honeycutt, M.E., A.S. Jarvis, and A.A. McFarland (1996) Cytotoxicity and mutagenicity of 2,4,6-trinitrotoluene and its metabolites, Ecotoxicol. Environ. Saf., 35, 282-287.

Isbister, J.D., G.L. Anspach, J.F. Kitchens, and R.C. Doyle. (1984) Composting for decontamination of soils containing explosives, Microbiologica 7, 47-73.

Kaplan, D.L. and A. M. Kaplan (1982) Thermophilic biotransformations of 2,4,6-trintrotoluene under simulated composting conditions. Appl. Environ. Microbiol. 44, 757-760.

Knicker, H., D. Bruns-Nagel, O. Drzyzga, E. v. Löw, and K. Steinbach (1999) Characterization of ^{15}N-TNT residues after an anaerobic/aerobic treatment of soil/molasses mixtures by solid-state 15N NMR spectroscopy, I. Determination and optimisation of relevant NMR spectroscopic parameters, Environ. Sci. Technol. 33, 343-349.

Michels, J. and G. Gottschalk (1995) Pathway of 2,4,6-trinitrotoluene (TNT) degradation by *Phanerochaete chrysosporium*, in Biodegradation of nitroaromatic compounds, Ed. J. Spain, Plenum Press, New York, USA.

Neumann, H.-G. (1996) Toxic equivalence factors, problems and limitations, Food Chem Toxicol, 34, 1045-1051.

Tan, E., C.H. Ho, W.H. Griest, and R.L. Tyndall (1992) Mutagenicity of trinitrotoluene and its metabolites formed during composting, J. Toxicol. Environ. Health. 36, 165-172.

Thorn, K. A., P. J. Pettigrew, W. S. Goldberg, E. J. Weber (1996) Covalent binding of aniline to humic substances. 2. ^{15}N NMR studies of nucleophilic addition reactions. Environ. Sci. Technol. 30, 2764-2775

Tweedy, B.G., C. Loeppky, and J.A. Ross (1970) Metobromuron: acetylation of the aniline moiety as a detoxification mechanism, Science, 168: 482-483.

Winterberg, R., E. von Löw, T. Held (1998) Dynamisches Mietenverfahren zur Sanierung von Rüstungsaltlasten, TerraTech, 3, 39-41.

PART 5
PHYTOREMEDIATION

PHYTOREMEDIATION

THOMAS MACEK, MARTINA MACKOVA,
PETRA KUCEROVA, LUDMILA CHROMA, JIRI BURKHARD
AND KATERINA DEMNEROVA

TM: Institute of Organic Chemistry and Biochemistry, Academy of Sciences of the Czech Republic, Flemingovo n. 2, 166 10 Prague, E-mail: tom.macek@uochb.cas.cz, fax: (+420-2)-24310090, MM, PK, LC & KD: Department of Biochemistry and Microbiology, Faculty of Food and Biochemical Technology, Institute of Chemical Technology, Prague, Technicka 3, 166 28 Prague; JB: Faculty of Environmental Chemistry, Institute of Chemical Technology, Prague, Technicka 3, 166 28 Prague, Czech Republic

Summary

Phytoremediation is the direct use of living green plants to degrade, contain or render harmless various contaminants of the environment, including recalcitrant organic compounds or heavy metals. The methods involved include phytoextraction, direct phytodegradation, rhizofiltration, phytovolatilisation or formation of artificial wetlands and lagoon systems. More research background and development of plants tailored for remediation needs, using genetic engineering and deeper understanding of plant co-operation with microorganisms is needed. The approach is illustrated by our studies of heavy metal uptake improvement or studies of the PCB conversion, which include screening of plant species *in vitro*, analysis of the products, evaluation of their toxicity and field plots, but also studies of enzymes and cloning of foreign genes into plants.

1. Introduction

1.1. GENERAL INTRODUCTION

There is no doubt about the fact that our environment is much more polluted than acceptable. Many toxic compounds are thus entering the food chain, endangering the biodiversity and decrease the chances for sustainable development. Removal of old contamination or such resulting of recent accidents is an everyday necessity. Due to high expenses connected with the use of classical physical or chemical methods for the

S.N. Agathos and W. Reineke (eds.).
Biotechnology for the Environment: Soil Remediation, 115-137.
© 2002 Kluwer Academic Publishers. Printed in the Netherlands.

today. It uses bacteria and fungi, and recently also plants for removal of contaminants from soil, water or sediments. Phytoremediation can be defined as the direct use of living green plants to degrade, contain or render harmless various contaminants of the environment (Cunningham *et al.* 1995), pollutants and xenobiotics, including heavy metals, radioactive elements or recalcitrant organic compounds. In an optimal case, the approach might lead to mineralisation of the organic compounds, with main aim to prevent migration of pollutants to a site of actual danger to human health (Salt *et al.* 1995). Much interest is devoted to explosives (French *et al.* 1999) and CWD (chemical weapons demilitarisation) (Macek *et al.* 1998, 1999). Recently also endocrine-disrupting chemicals (EDC) like tributyl tin and bisphenol A are coming to the centre of interest. Special attention should be paid to the application of *in vitro* systems for basic research on the role of plants for the remediation of contaminated sites or flows, and in the improvement of their effectiveness.

To render inoffensive harmful compounds, we need also to consider what products are formed, and also what is their toxicity towards plants, animals or man, and how these compounds will be further metabolised by soil microorganisms (Macek *et al.* 2000a). This is an aspect, which was sometimes omitted, in practical applications due to the struggle for fast and inexpensive outcomes of remediation processes.

In contrast to stripping the contaminants from the soil using physical, chemical or thermal processes, transfer of soil to hazardous waste landfills etc., an *in situ* approach like phytoremediation has additional advantages. The establishment of vegetation on a contaminated site also reduces soil erosion by wind and water, which helps to prevent the spread of contaminants and reduces exposure of humans and animals. Olson and Fletcher (2000a) describe ecological recovery of vegetation at a former industrial sludge basin, together with its implications for phytoremediation.

Considering the rapidly growing world population and the detrimental impact of agricultural systems on the environment (Zechendorf 1999), cleaning of large contaminated sites is important for ensuring sustainable development, because it might reduce pressure to expand into wilderness, rain forests and marginal lands, thus supporting biodiversity and preservation of vital ecosystems.

1.2. PHYTOREMEDIATION AND RHIZOREMEDIATION

1.2.1. Metabolic aspects of phytoremediation

Plants remove organic compounds from soil by direct uptake of the contaminants, as summarised by Schnoor *et al.* (1995), followed by subsequent transformation, transport and their accumulation in a non-phytotoxic form, which does not necessarily mean non-toxic for humans. Basically the plants show some similarities with the mammalian liver by their ability to metabolise a wide spectrum of xenobiotics and by the pathways involved. For this reason plants are sometimes called the green liver of the planet - organic compounds are metabolised by plants in a similar manner as in liver, in three phases including activation and conjugation of lipophilic molecules with hydrophilic residues. In the third phase polar products are stored in vacuoles, or bound into insoluble cell structures by oxidative polymerisation instead of being excreted.

If the biotransformation reactions lead to decrease of the toxicity of a given xenobiotic compound, the reaction is called detoxification. In some cases the biotransformation reactions in general can lead also to an increase of the toxicity of the compounds. In such a case we speak about activation of xenobiotics, especially important in the case of metabolic activation of non-mutagenic compounds (promutagens) into mutagenic compounds. This field is not included in this chapter, but it is discussed in detail in the review by Plewa and Wagner (1992).

Different groups of enzymes were proposed to play an important role in detoxification. Concerning the first step e.g. microsomal cytochromes P450 were proved to oxidise xenobiotics, but in the same compartment also peroxidases can be found, and perform reactions with similar outcome (Stiborová and Anzenbacher 1991). Peroxidases are ubiquitous in plants, and their activity can be correlated with the conversion of some xenobiotics by plant cells (Kucerová et al. 1999). Enzymes involved in xenobiotic metabolism were reviewed by Sandermann (1992, 1994). He showed that cytochrome P450, peroxygenases, and peroxidases are involved in plant oxidation of xenobiotics. Other enzyme classes like glutathione S-transferases, carboxylesterases, O-glucosyltransferases O-malonyltransferases, N-glucosyltransferases and N-malonyltransferases are associated with xenobiotic metabolism in plant cells, transport of intermediates and compartmentation processes. The role of glutathione-mediated detoxification systems in plants was recently discussed by Dixon et al. (1998).

1.2.2. Rhizoremediation

The living plant roots exert strong changes in the physical, chemical and biological properties of the soil. The complexities of roots and their rhizosphere were discussed recently by McCully (1999). The soil-root interface, rhizoplane, and the narrow volume surrounding roots (a few mm) called the rhizosphere, is characterised by several processes such as exudation of organic compounds, root respiration (absorption of oxygen and release of carbon dioxide), release of protons and other mineral ions, and uptake of water and solutes. The influence of plant roots on the soil environment is widely discussed by Morel et al. (1999). Various groups of microorganisms are present in the soil (bacteria, actinomycetes, fungi, algae, and viruses). In the rhizosphere their numbers are higher than in the bulk soil. The ratio of microorganisms in rhizosphere and in the soil (the R/S ratio) varies from 2 to 20, in some cases reaching 100. Qualitative changes of microbial population were also observed (Fletcher et al. 1995, Morel et al. 1999). Composition of microbial population is controlled both by soil factors and by plant factors, including compounds with allelopathic effects. The intensive microbial activity in the rhizosphere is due to the presence of high amounts of available carbon released as exudates by roots. The effects of plant compounds on microbial activity were studied by Donelly et al. (1994). Hedge and Fletcher (1996) showed the influence of growth stage and season on the release of root phenolics by mulberry as related to the development of phytoremediation technology. Olson and Fletcher (1999) evaluated in field the mulberry root structure with regard to phytoremediation. In addition to the function of carbon source some of the compounds released by plants can serve as inducers of the degradation pathways, as described by

Fletcher *et al.* (1995) or Gilbert and Crowley (1997) using different bacterial PCB-degraders.

Plants respond to the presence of microorganisms by modification in growth, e.g. symbiosis or interactions with free living organisms. Rhizobium-legumes or endo- and ectomycorhizas can serve as examples of direct and positive actions (Batkhuugyin *et al.* 2000), while pathogenicity is a negative effect. Donelly and Fletcher (1994) studied potential use of mycorhizal fungi as bioremediation agents. Indirect actions involve growth promoting substances, antibiotics or siderophores released. Positive effects are expected from inoculation of roots by selected microorganisms (suppliers of growth-promoting substances, nitrogen fixation, acquisition of phosphorus etc.). Such an example is the finding of a plant growth-promoting bacterium that decreases nickel toxicity in seedlings. Burd *et al.* (1998) described *Kluyvera ascorbata* SUD165, which produced a siderophore, and displayed 1-aminocyclopropane-1-carboxylic acid deaminase activity. The presence of the bacterium did not reduce the nickel uptake by seedlings in contaminated soil, thus probably promoting the plant growth by lowering the level of the plant stress hormone ethylene that is induced by nickel.

The field release of bioluminescent *Sinorhizobium meliloti* is in this respect very interesting. A strain of this organism is prepared by Tebbe *et al.* (1998), using genetic engineering with inserted monitoring genes. This technique allows to detect such cells specifically and easily when isolated from environmental samples.

1.2.3. Exudates and enzymes released

Root exudation is the major process associated with the rhizosphere. In addition to uptake and direct phytodegradation plants support bioremediation by release of exudate and enzymes that stimulate both microbial and biochemical activity in the surrounding soil and mineralisation in the rhizosphere. The composition of exudates, sites of exudation and various factors affecting the root exudation are summarised by Morel *et al.* (1999). These include nutritional stress, excess of metals or presence of some microorganisms. Exudates include compounds of low molecular weight, like phenolics or sugars and amino acids and other compounds, secretions, lysates, and also compounds of high molecular weight, i.e. mucilage. Concerning the enzymes in the exudate, it was proven by immunological methods that these enzymes in soil are of plant origin, and have wide substrate specificity. The enzymes involved, include peroxidases, nitroreductase (nitroaromatic compounds, 2,4,6-TNT), dehalogenases (chlorinated solvents, hexachloroethane), laccase (anilines, 2,4,6-triaminotoluene) and nitrilases.

1.3. METHODS USED IN PHYTOREMEDIATION

The diverse approaches include phytoextraction, direct phytodegradation, rhizofil-tration, and formation of artificial wetlands and lagoon systems, co-operation with microorganisms in the process of rhizoremediation, or development of plants tailored for specific phytoremediation needs, using genetic engineering.

To illustrate the complexity of the task, our studies concerning heavy metal uptake (Macek *et al.* 2000c) and its improvement by genetic engineering, and especially studies on the conversion of PCBs by plants are presented (Macek *et al.* 2000b), including the screening of different plant species *in vitro*, analysis of the products, studies of enzymes involved, cloning of bacterial genes, studies on plant secondary metabolite support to rhizospheric bacteria, pot tests in real contaminated soil and field plots.

Phytoremediation is favourably used in places with surface contamination (unto 5 meter depth) and it was found that it is highly effective also for hydrophobic pollutants, such as benzene, toluene, ethylbenzene, xylene, chlorinated xenobiotics, nitro-compounds or nitrotoluene compounds. It has a great potential for application in remediation of residues from petrochemical production, chemical industry dump sites, in places of ammunition waste storage, in places with contamination by pesticide residues, but also municipal wastewater or other effluents. Phytoremediation is discussed in connection with natural attenuation by Olson and Fletcher (2000b).

1.3.1. Phytostabilisation

Phytostabilisation helps to stabilise the soil surface by covering the contaminated soils with adapted plants thus reducing the risk of transport of pollutants adsorbed on fine solid phase to water streams. It also reduces or prevents wind erosion. Also, growing tolerant plants reduces the water movement into the soil profile, limiting the leaching of soluble pollutants. This technique can be used to re-establish a vegetative cover at sites where natural vegetation is lacking e.g. due to high metal concentration in surface soil or physical disturbances to surface materials.

1.3.2. Phytoextraction

Phytoextraction (also called phytoaccumulation) of heavy metals is based on the ability of plants to take up, translocate and concentrate some substances in their tissues, as well in root as in upper parts. Approximately 400 plant species have been classified as hyperaccumulators of heavy metals, and most of them accumulate Ni (Baker *et al.* 1994). The phytoextraction scenario consists of cultivating hyperaccumulating plants on the contaminated soil and harvesting the above ground plant material (leaves and stalks) which would than be dried or incinerated and disposed off safely in a landfill or even processed for metal recovery (Morel *et al.* 1999). This procedure may be repeated as necessary to bring soil contaminant levels down to allowable limits.

1.3.3. Rhizofiltration

Rhizofiltration is the adsorption or precipitation onto plant roots or absorption into the roots of contaminants that are in solution surrounding the root zone. Rhizofiltration is similar to phytoextraction, but the plants are used primarily to address contaminated groundwater rather than soil. According to an EPA Technology Fact Sheet (Anonymous 1998), the plants to be used for cleanup are raised in greenhouses with their roots in water rather than in soil. Plants are then acclimated in contaminated water and after that planted in the contaminated area.

1.3.4. Phytodegradation

Phytodegradation, also called phytotransformation, is the breakdown of contaminants taken up by plants through metabolic processes within the plant, or the breakdown of contaminants external to the plants through the effect of enzymes produced by the plants (see parts 1.2.1., 1.2.3.).

1.3.5. Phytovolatilisation

Phytovolatilisation is the uptake and transpiration of a contaminant by a plant, with release of the contaminant or a modified form of it to the atmosphere from the plant. The effect occurs as growing trees and other plants take up water and the organic contaminants. Some of these compounds can pass through the plants to the leaves and evaporate into the atmosphere. As shown for poplar trees, 90 % of TCE (trichloroethylene) taken up by the tree can be volatilised. Plants can also extract volatile elements (e.g. mercury and selenium) and volatilise them from the leaves. It is open to discussion, of course, if it is a good solution to transfer the problem from water to air. Some aspects of root uptake of tetrachloroethylene were addressed using tomato plants, showing high level of bound contaminant (Dannel *et al.* 1995).

1.3.6. Hydraulic control

Hydraulic control of contaminants appears when plants act as hydraulic pumps. In case their roots reach down toward the water table and establish a dense root mass that takes up large quantities of water, poplars can transpire between 50 and 300 gallons of water per day out of the ground. There are several applications that use plants for this purpose, such as riparian corridors of plants applied along a stream or river, buffer strips applied around the perimeter of landfills (Anonymous 1998).

1.4. ARTIFICIAL WETLANDS

The use of wetlands for treatment of contaminated water is well established and cost effective. Many such systems exist around the world, in most cases treating wastewater for nitrogen, phosphate and biological oxygen demand (BOD) originating from municipal works, farming and small industries. The majority of treatment wetlands rely on a combination of moderately reducing substrate conditions and the oxidising conditions surrounding plant roots (Hammer 1989). For several types of wastewater, wetlands with strongly reducing substrates, facilitating reduction of sulphate are also considered. Most treatment wetlands in existence are surface flow wetlands. The effect of wetland plants on the biogeochemistry of metals beyond the rhizosphere was addressed by Wright and Otte (1999).

The wetland systems have to be mentioned when discussing phytoremediation, because the wetland plants are an inherent component of engineered wetland systems and are believed to play a pivotal role in the removal process in these systems. The most commonly used plant species in treatment wetlands are the monocotyledon species common reed (*Phragmites australis*), cattail (*Scirpus lacustris*), but also sweetgrass (*Glyceria fluitans*) is considered. The removal mechanisms and types of constructed wetlands are described by Vymazal *et al.* (1998). An interesting technological approach is that of ponds with a floating plant cover. The plant cover reduces wave formation and as a consequence oxygen input into the water body is restricted.

Furthermore the plants, as in surface flow systems, play an important role as a carbon source (decaying plant matter and root exudates) for microbial anaerobic processes. The plant roots also provide a surface for optimal growth of biofilms. At present, knowledge is lacking about the fundamental processes in such systems and especially metal fixation rates are not available. O'Sulivan *et al.* (1999) discussed recently the use of wetlands for rehabilitation of metal containing mine wastes.

1.5. PERSPECTIVES OF PLANTS IN DETOXIFICATION IN CWD

Army officials show an increasing interest in phytoremediation technologies, as was apparent at different NATO workshops. There are already a few possibilities of the exploitation of plants discussed from the chemical weapon demilitarisation (CWD) point of view. As proposed by Macek *et al.* (1998), plants can be used for detoxification of some chemical warfare compounds as such, and also for the detoxification of some products formed during breakdown or neutralisation of chemical warfare agents in demilitarisation processes. Problems are arising from incomplete reactions, from products themselves and also the soil contaminated by leaking of corroded munitions represents a specific problem. Plants are often considerd to be only useful for the final polishing of sites decontaminated in other ways, but it was shown that plants and their enzymes might play an active role in degradation of many toxic compounds directly as part of the technological set-up (Macek *et al.* 1998, 1999). The possibility to remove products of yperite breakdown from soil was studied recently by Zakharova *et al.* (2000). The breakdown products were mixed with soil (2g/kg) and seeds of various plants were potted into that soil. The sulphur containing products were effectively taken up from contaminated soil by plants.

Of special interest are the abilities of plants to metabolise explosives. Some water plants (Hughes *et al.* 1997) have been described to convert TNT to aminodinitrotoluene (unfortunately more toxic than TNT) and some minor products, while some plants are able to degrade nitro-glycerine, cyclotrimethylenetrinitramine (RDX) and other compounds. Many research groups are studying this particular field (Hubalek and Vanek 1998, Bhadra *et al.* 1999), because there are quite many sites with soil containing high amounts of explosives around the world (Nepovím and Vanek, 2000). In this field there are great expectations connected with transgenic plants (French *et al.* 1999).

For treatment of effluents from CWD technologies, artificial wetlands and lagoons are also under consideration, connecting the activities of water plants and bacteria, allowing larger volumes to be decontaminated under continuous conditions.

2. Examples of practical phytoremediation experiments

Not all waste types and site conditions are comparable. Each site must be individually investigated and tested. Engineering and scientific judgement must be used to determine if a technology is appropriate for a site, as EPA repeatedly states (Anonymous 1998). Researchers are finding that the use of trees rather than smaller plants allows them to treat deeper contamination because tree roots penetrate more deeply into the ground. Very deep layers of contaminated groundwater can be treated by pumping the water and using plants to treat the contamination.

2.1. EXAMPLES OF ORGANIC POLLUTANTS AND XENOBIOTICS REMOVAL

It is already experimentally proven that degradation of some organic compounds is achieved much faster in soils with than without vegetation cover, as examples can serve e.g. crude oil alkanes, polyaromatic hydrocarbons (PAH), discussed by Yateem *et al.* (1999) or Pradhan *et al.* (1998), chlorinated pesticides (PCP, 2,4-D), other chlorinated compounds (PCB, TCE), discussed by Susarla *et al.* (1999), explosives (TNT, DNT, RDX), organophosphorous insecticides, surfactants (detergents).

An example of successful treatment among the limited number of full-scale applications is the use of hybrid poplar trees on one acre in Maryland, at a U.S. Army testing facility, over a shallow groundwater plume contaminated with organics from several toxic disposal pits. It was proven that poplar trees acting as hydraulic pumps prevent spread of contaminants to a nearby marsh, and treat the contaminated plume by phytovolatilisation and rhizofiltration. A recent summary by the US EPA (Anonymous 1998) shows the successful application of phytoremediation. In New Gretna, NJ, phytodegradation of chlorinated solvents was described using hybrid poplars; in Ogden, UT, phytoextraction and rhizodegradation of petroleum hydrocarbons from soil and groundwater using alfalfa, poplar, juniper and fescue; in Portsmouth, VA, phytodegradation of petroleum by grasses and clover; in Portland, OR, phytodegradation of PCP and PAHs from soil by ryegrass; in Milan, TN, phytodegradation of explosives wastes from groundwater was proven to be performed by duckweed and parrotfeather; in Texas City, TX, mulberry was tested as vegetative cover with rhizodegradation of PAHs. These examples, just some of the ongoing field tests, allow for optimism concerning the future of phytoremediation in the field of organic compounds.

2.2. EXAMPLES OF HEAVY METAL REMOVAL

The fact that some plants are resistant to high concentrations of heavy metals was known since long to prospectors looking for metal ores. Some plants can accumulate surprising amounts of some metals (Brown *et al.* 1995), probably by mechanisms similar to the uptake of metals essential for their enzymatic activities. It seems that these so-called hyperaccumulators have a selective advantage when growing on high-metal containing soil, being protected e.g. against herbivores. Unfortunately, it seems that most of the known hyperaccumulators show low biomass yield, while most of the high productive plants accumulate too low amounts of heavy metals to represent an economically feasible approach for cleaning of contaminated soil.

Metal concentrations in shoots of some known hyperaccumulators can reach high levels of heavy metals as summarised in Table 1.

Examples of the application of phytoremediation in the field are hydroponically grown sunflower in artificial lagoons on a river near Chernobyl and at a uranium-enrichment plant in Ohio (which culture was absorbing radioactive metals), or phytoextraction of heavy metals and radionuclides from soil by Indian mustard in Trenton, NJ (Salt *et al.* 1995).

Table 1: *Heavy metal content in hyperaccumulating plants after Cunningham and Ow (1996).*

Element	Species	Level
		(on a dry weight basis).
Mn	*Macadamia neurophylla*	51 800 mg kg^{-1}
Zn	*Thlaspi caerulescens*	51 600 mg kg^{-1}
Ni	*Psychotria douarrei*	47 500 mg kg^{-1}
Cu	*Ipomoea alpina*	12 300 mg kg^{-1}
Co	*Haumaniastrum robertii*	10 200 mg kg^{-1}
Pb	*Thlaspi rotundifolium*	8 200 mg kg^{-1}
Cd	*Thlaspi caerulescens*	1 800 mg kg^{-1}

Kayser and Felix (1998) evaluated the efficiency of the use of metal accumulating plants based on long-term (5 years) field trials at three heavy metal contaminated sites in NW Switzerland. They concluded that the long time required for significant reduction of toxic metal content in soil is the limiting factor for wider exploitation of plants for this purpose. Nevertheless, phytoextraction of metals presents large economic opportunities because of the size and scope of environmental problems associated with metal-contaminated soils and the competitive advantages offered by a plant-based technology.

2.3. ADVANTAGES AND ECONOMICAL ASPECTS

The main advantages of phytoremediation compared to other methods are:

- it is far less disruptive for the environment,

- there is no need for soil disposal sites,

- it has better public acceptance,

- it avoids excavation and heavy traffic,

- it allows much larger scale clean up.

The American Society for Plant Physiology (ASPP) estimated a few years ago, that conventional treatment of contaminated soil costs as much as USD 1000 per tonne of soil, while phytoremediation about USD 30 per tonne of soil (Jones 1994). These numbers are just rough averages, with great differences depending on pollutant type, soil type, and size of the contaminated site and many other factors. Recently several authors have summarised the financial aspects of phytoremediation. In each case the phytoremediation was much cheaper when compared to any other approach. On sites where the contamination is spread over a large area, phytoremediation is the only economically feasible approach (Jones 1994), (Salt *et al.* 1995).

Specifically, two subsets of phytoremediation are nearing commercialisation. The first-one is phytoextraction, in which high biomass metal-accumulating plants and appropriate soil amendments are used to transport and concentrate metals from the soil into the harvestable part of roots and above-ground shoots, which are harvested with conventional agricultural methods. The second-one is rhizofiltration, in which plant roots grown in water absorb, concentrate and precipitate toxic metals and organics from polluted effluents.

According to a survey of literature, which includes several claimed cost estimations not necessarily substantiated by the data, the consensus cost of phytoremediation has been estimated at $25 to $100 per tonne of soil treatment and $ 0.60 to $6.0 per 1 000 gallons for treatment of aqueous waste-streams. In both cases, the cost for remediation of organic contaminants can be expected to fall be at the lower end of these ranges and the cost for remediation of heavy metals to be at the higher end (Boyajian and Carreira 1997). The expenses of phytoremediation are in both application domains much less than the half the cost for any other equally effective treatment. According to 1997 estimates by the US EPA, the cost of using phytoremediation in the form of alternative cover (vegetative cap) ranges from $10 000 to $30 000 per acre, which is said to be two- to five-fold less expensive than traditional capping (Flathman *et al.* 1999).

3. Basic research aspects

3.1. PLANT *IN VITRO* CULTURES IN PHYTOREMEDIATION STUDIES

Most of the experiments used for establishing the phytoremediation techniques have been done with normal soil-grown plants or by hydroponical culture. Nevertheless, as more and more effort is drawn toward basic research to understand and improve the performance of plants in remediation technologies, the number of results obtained with the help of *in vitro* plant cell, tissue and organ cultures is rapidly increasing. It was proven that such cultures are able to metabolise organic compounds by common metabolic pathways and can be exploited as a useful system with some advantages in comparison with intact plants. The plant material used includes homogeneous non-differentiated callus and suspension cultures, differentiated embryogenic cultures, shooty teratomas and hairy root cultures obtained by genetic transformation by the R*i* plasmid using *Agrobacterium rhizogenes* or cultures transformed by the TI (tumour inducing) plasmid of *A. tumefaciens*. These soil bacteria are the most often used vectors for transferring foreign genes into plants, but in this case wild type bacteria were used without modification by genetic engineering.

All the materials mentioned can be grown under standard laboratory conditions, the growth is independent of the weather, plant *in vitro* cultures grow more rapidly than normal plants and the results can be obtained with less analytical expense.

3.1.2. Callus and cell suspension cultures

The number of results obtained with the help of *in vitro* plant cell cultures is rapidly increasing. The concept is not new; *in vitro* cultivated plant cells have been used in studies of herbicide resistance and metabolism for many years (Harms and Kottutz

1990, Komosa *et al.* 1992, Komosa and Sandermann 1992), or pesticide metabolism (Lamoureux and Frear 1979, Wimmer *et al.* 1987, Vanek *et al.* 1989a, b). Also other organic xenobiotics were studied, e.g. metabolism of pentachlorophenol by wheat and soybean cell suspension cultures (Langebartels and Harms 1984). The metabolism of PCBs by plant cells was studied already 10 years ago by Groeger and Fletcher (1988). Metabolism of many other organic compounds has been studied with the help of such plant models, e.g. the metabolism of sulphonated aromatic compounds by rhubarb cells (Duc *et al.* 1999a, b).

Cultures proved to be a very useful tool, but the results obtained under *in vitro* conditions have to be carefully evaluated, especially in the case of non-differentiated callus and suspension cultures. We screened over 40 cultures of 14 plant species for their ability to metabolise PCBs, to find models for effective studies of product formation and enzymes involved. It was found that the variability within more cultures of the same species is high (Macková *et al.* 1997a). Among twelve *in vitro* cultures of the same species (*Solanum aviculare* FORST), there were found some strains not converting PCBs at all, while others exhibited very high conversion rates under the same conditions. This was studied on stable cultures established more than a decade ago by Macek (1989). The PCB degrading activity depends much on the level of morphological and biochemical differentiation of the strain or clone tested. For this reason different plant species cannot be evaluated and their performance compared on the basis of analysis of a single *in vitro* culture of each species, as did Wilken *et al.* (1996), suggesting the best species out of 12 tested on the base of comparison of single suspension cultures. On the other hand, comparison of more cultures shows the trend which allows to predict the usefulness of a given species, and independently of the "negative" clones where some enzyme may be missing, those which detoxify, convert or degrade, are a very important tool in establishing the fate of the xenobiotic in the plant cell.

Callus cultures have been used also in work with heavy metals, for selection and biochemical characterisation of zinc and manganese adapted lines in *Brassica* spp. (Rout *et al.* 1999). Metal (Hg, Cu, Cd, Zn) stress response and tolerance in connection with heat stress proteins were studied e.g. using cell cultures of *Silene vulgaris* and *Lycopersicon peruvianum* by Wollgiehn and Neumann (1999).

3.1.3. Hairy root cultures

The so-called "hairy root" cultures are formed by genetic transformation of a single plant cell by the soil bacterium *Agrobacterium rhizogenes*. Due to their fast growth, unlimited propagation in culture media, genetic and biochemical stability and growth in hormone-free media these tissues proved to be very good model systems for plant metabolism and physiology. Doran (1997) published the only volume, until now, dedicated entirely to hairy root cultures, and their applications are spreading fast since, as summarised in a review by Shanks and Morgan (1999).

The exploitation of transformed hairy root cultures in phytoremediation is especially rewarding, as discussed by Macková *et al.* (1997a, b) in PCB degradation studies, or as described by Betts (1997). Much has been done in studies on the degradation of explosives by Hughes *et al.* (1997). This is because *Agrobacterium rhizogenes* transformed plant roots exhibit nearly all features of normal plant roots and grow rapidly under defined aseptic conditions *in vitro*, thus allowing to distinguish between the plant metabolism itself and the effect of the complex interaction between plants and microbial communities in the rhizosphere. Bhadra *et al.* (1999), used such a model for confirmation of conjugation processes during TNT metabolism by axenic plant roots, presenting rigorous mass balance data and a comprehensive set of analytical data for TNT transformation in plant tissues, in the first report of the types of conjugates formed from this widespread and problematic contaminant. The growing interest in exploitation of hairy root models is summarised in two recent review articles by Shanks and Morgan (1999) and Pletsch *et al.* (1999).

The system is especially useful for the study of the internal fate of xenobiotics taken up by the plan. The limitation is in altered properties of the *in vitro* cultivated hairy root surface compared to the roots (or even hairy roots) grown *in vivo* with full production of cuticle, waxes and other compounds that influence the sorption and uptake of xenobiotics. Nevertheless, hairy roots play an important role also in heavy metal uptake studies, as shown e.g. by Macek *et al.* (1994, 1997a, b) in studies of cadmium uptake, or Soudek *et al.* (1998a, b) comparing the uptake of different metals by horse radish hairy roots.

3.2. HEAVY METALS

Plant systems are capable of concentrating some toxic inorganics, as described under heading 2.2. of this chapter. Finding or developing plants that acquire high levels of metal contaminants in harvestable tissue was thought impossible until the (re)discovery of a small group of remarkable plants called hyperaccumulators. These plants are relatively uncommon, but present throughout in the plant kingdom. Harvesting plants, on the other hand, is a familiar agricultural technology (Salt *et al.* 1995).

The most important parameter for selection of suitable plants is not the tolerance of the plant to heavy metals, but the effectiveness in the accumulation of heavy metals. In addition to accumulation capacity, biomass production must be considered in order to determine the total metal uptake. The tolerance to some heavy metals was recently studied by Bringezu *et al.* (1999). In the case of contaminated water a wider range of organisms seems to be promising, from water plants to microalgae, from root filters to immobilised bacteria. Waste products such as sludge, ash, wastewater, mine drainage water and urban refuse city dumps have high nutrient values but often also a high heavy metal content. Rhizofiltration, in which plant roots grown in water, absorb, concentrate and precipitate toxic metals from polluted effluents is tested.

The contaminant must be in a biologically accessible form, and root absorption must take place. Translocation of the contaminant from root to shoot makes tissue harvesting easier and lessens exposure to the contaminant. Decontamination of a site requires a plant with both a high biomass yield and high metal accumulation of 1 to 3 % of the metal on dry matter. Although some of the properties desired for phytoremediation are present in currently recognised hyperaccumulators, many of these plants have low biomass productivity and restricted element selectivity in hyperaccumulation. In addition, there is little or no information available about the agronomics, genetics, breeding potential and disease spectrum of most of these plants.

The response of higher plants to cadmium in their environment is summarised in much detail by Sanita di Toppi and Gabbrielli (1999). They discuss many factors, like sequestration, cell wall immobilisation, plasma membrane exclusion, stress responses etc., explaining models of tolerance and detoxification mechanisms. The rapid accumulation of technetium in plants and proteins that bind it are described by Krijger *et al.* (1999).

After harvesting, a processing step is necessary. Alternatively, the harvested biomass could be reduced in volume and/or weight by composting, anaerobic digestion, low temperature incineration, leaching. This step would further decrease the costs of handling and processing, as discussed by Salt *et al.* (1995). After stabilisation potential subsequent landfilling can be considered. With some metals, after leaching or smelting in kiln (e.g. Ni, Zn, and Cu), the value of the reclaimed metal may provide an additional incentive for remediation. This step reduces generation of hazardous waste and generates recycling revenues.

Some chemical war agents (CW agents or CWs) like Adamsite (diphenylamine-chloroarsine, Chem. Abstr. reg. no. 578-94-9) or Clark (Clark1 is diphenylchloroarsine, CA reg. no. 712-48-1, Clark2 is diphenylcyanoarsine, CA reg. no. 23525-22-6) contain a high percentage of arsenic, which requires special treatment. It is known that plants are able to take up and concentrate not only metals but also arsenic (Tlustos *et al.* 1997)

The compounds which plants use for binding of heavy metals were recently reviewed by Kotrba *et al.* (1999). In general there is still an important relative lack of knowledge of heavy metal transport, vacuolar uptake, ABC transporters and other factors important for a really effective exploitation of plants in heavy metal remediation.

3.3. METABOLISM OF PCBs BY PLANTS

In the centre of our interest (Macková *et al.* 1998) are polychlorinated biphenyls (PCB), which are among the most recalcitrant pollutants in the world. They have been used in developed countries since 1930 in the production of electrical transformers, capacitors, in hydraulic systems and gas turbines, vacuum pumps, in paints as fire retardants and plasticizers, mainly due to their excellent chemical, physical, and biological stability. PCBs have been distributed all over the globe even into remote places such as Antarctic. Although the manufacture of PCBs has already been prohibited in many countries as early as 1970 (but in the former Czechoslovakia the production was only stopped in 1984), they are still an environmental problem. These days there is, for instance in the Czech Republic, no effective technology available, for their efficient removal. Thus, bioremediation through the use of specific biodegrading bacteria, plants and fungi is considered to be an effective approach for cleaning of PCBs contaminated sites (Káš *et al.* 1997, Macková *et al.* 1999).

This research is of high interest as the problem of PCB pollution is persistant and the cost of other kinds of PCB remediation is very high. The metabolic pathways of PCB conversion are different for bacteria, plants or fungi, thus various products are formed. In an environment where such organisms are present together (rhizosphere) the products can be further metabolised and influence the autochthonous fauna and flora. The ongoing research is aimed at increasing the general knowledge on the rhizosphere and particularly of PCB degradation in the soil.

The metabolism of PCBs by plant cells was addressed already by Groeger and Fletcher (1988), showing that plants are able to degrade PCBs. Lee and Fletcher (1992) suggested that cytochrome P450 is responsible for PCB conversion. In our studies we started with plant *in vitro* models cultivated for analytical method modification (Burkhard *et al.* 1997), for screening of cultures of different species (Mackova *et al.* 1997a, Macek *et al.* 1998) and optimisation of incubation conditions (Mackova *et al.* 1997b). On the basis of these results we could proceed to more profound investigation of the involved enzyme systems and of the product formed, as well as to toxicity studies, genetic engineering experiments and to pot and field studies in contaminated soil using normal plants.

We screened over 40 cultures of 14 plant species for their ability to metabolise PCBs. It was found that the variability within more cultures of the same species is high (Macková *et al.* 1997a). Best PCB conversion yields we obtained with the hairy root clone SNC-9O of *Solanum nigrum*, so this clone was used as the main model in further studies (Macková *et al.* 1997b, Kucerová *et al.* 1998, 2000).

3.3.1. The enzyme systems involved

Cole in his review (1983) showed that oxygenation is a common process, shared with other organisms and it is an important initial mode of attack when organisms encounter what are often highly lipophilic compounds. This step serves to increase water solubility and provides an opportunity for conjugation via glycosidic bond formation. Three phases of xenobiotic transformation in plants are metabolic sequences of transformation (phase I), conjugation (phase II) and compartmentation (phase III) reactions. Many early examples of such metabolic sequences in plants have been summarised and compared with animal metabolite patterns (Sandermann 1994). Similarities between animal and plant metabolic pathways exist but on the other hand plant metabolism may often be more complex and an important difference from animal metabolism appears to exist especially in the formation of residues (phase III).

Cytochrome P450 is suggested as the key enzyme responsible for activation of PCBs by Lee and Fletcher (1992). They studied the involvement of mixed function oxidase system in PCB metabolism. Our results indicate that peroxidases are also involved in the PCB metabolism in plants (Kucerová *et al.* 1998b, 1999). For this reason we performed a characterisation of the involvement of both enzymes in the PCB metabolism, which included estimation of enzyme activities (for peroxidases see Mackova *et al.* 2000a), isoenzyme pattern and inhibition analysis.

3.3.2. PCB transformation products

The products have to be studied, because it cannot be excluded that in some cases even more toxic compounds can be formed, or compounds more soluble and with better biological availability than the original ones. Recently it was shown that plants are able to metabolise PCBs, but the data for such models are scarce in comparison with the vast amount of papers on bacterial or fungal pathways of PCB degradation. Fletcher *et al.* (1987) showed transformation of 2-chlorobiphenyl but without description of the products obtained. In the beginning of the 1970s there were some studies discussing metabolism of PCBs in plants with the identification of oxygenated metabolites (hydroxychlorobiphenyls) confirmed by different analytical methods (Moza *et al.* 1974, 1976, Butler *et al.* 1992). The authors identified two monohydroxychlorobiphenyls and one dihydroxychloro derivative, which were mostly conjugated. They also found dechlorinated products. Wilken *et al.* (1995) stated that metabolism of defined PCBs congeners was dependent on plant species and they detected various monohydroxylated and dihydroxylated compounds after acid hydrolysis of polar metabolites.

The aim of our work (Macková *et al.* 1996) was the identification of products of the conversion of individual congeners, characterisation of hydroxylation pattern as a function of the position of the Cl atoms (Kucerová *et al.* 1998a) and analysis of the toxicity of PCBs and the products (Forman *et al.* 1998). The main products formed in the first step of transformation of monochlorobiphenyls and dichlorobiphenyls by plant cells were already identified (Kucerová *et al.* 2000) as hydroxylated chlorobiphenyls, and characterised by GC-MS comparison of silylated and nonsilylated samples with standards. The products derived from higher congeners are under investigation.

3.3.3. Pot and field experiments, rhizoremediation

The laboratory experiments have to be confirmed by field tests. For this reason our results have been checked by growing plants in real contaminated soil under field conditions. We used 10-l metal buckets to avoid sorption of PCBs into plastic, filled by soil from the contaminated sites, using regular watering. Different plant species selected on the basis of *in vitro* experiments were tested (Macková *et al.* 2000). Field plots have been established for three seasons, without addition of nutrients or watering. The results achieving about 20 % PCB removal during one season are summarised by Macková *et al.* (2000b). In these experiments also the co-operation of plants and bacteria in rhizosphere is being studied. It was proven that plants support the growth of rhizospheric bacteria in comparison with non-vegetated controls.

3.3.4. Improvement of plant material for PCB degradation

We started research in the field of genetic engineering with the aim to improve purposefully the abilities of plants for this purpose (Macek *et al.* 1998, Borovka *et al.* 1998) by introducing of the gene *bphC* of the bacterial biphenyl operon, responsible for ring cleavage of the biphenyl metabolite, 2-3 dihydroxybiphenyl. This is the reaction step, which is missing in plants, while some bacteria are able to cleave the PCB-ring. Different aspects of our research are presented in the following paragraphs.

4. Genetic modifications

4.1. BREEDING AND GENETIC ENGINEERING

Enhancement of metabolic abilities of plants by can be achieved e.g. by traditional breeding, protoplast fusion and direct insertion of novel genes. The methods of genetic engineering are widely used for the improvement of different crop plants. The introduced genes usually bring higher resistance against pests or herbicides, or improve the technological properties of the plant. This is especially remarkable in the field of herbicide resistance, but also genes responsible for other important traits are being introduced into plant species. A similar approach is expected to largely improve the plant abilities in the field of environmental detoxification (Macek *et al.* 1997b).

There are many attempts to breed willow, poplar and other plants with properties useful for phytoremediation. Very promising results were obtained by means of genetic engineering. The aim is the formation of plants combining high ability to accumulate, detoxify or degrade xenobiotics and pollutants, with resistance towards the toxic compounds present and with suitable agronomic characteristics, in other words the improvement of the process by using genetically modified plants specifically tailored for the phytoremediation purposes

At present there are two main goals: 1) metal-hyperaccumulation traits might be introduced into fast growing, high-biomass plants or vice versa, and 2) transgenic plants can harbour microbial genes for biodegradation of organic compounds.

4.2. IMPROVEMENT OF THE DEGRADATION OF ORGANIC COMPOUNDS

Some examples are already available to illustrate the degradation of organic pollutants and xenobiotics. Degradation of explosives was the target of a recent study from Cambridge, UK (French *et al.* 1999). The starting point was that plants denitrating glyceroltrinitrate had already been described, that some aquatic plants degrade TNT and that it is known that *Enterobacter cloacae* PB2 utilises pentaerythritol tetranitrate. The authors succeeded in preparing transgenic plants expressing pentaerythritol tetranitrate reductase, which were characterised by better germination and growth on agar media containing explosives, and showed enhanced degradation of PETN (pentaerythritol tetranitrate), GTN (glyceral trinitrate, nitroglycerin), TNT (trinitrotduene). The results are not yet confirmed by tests in soil.

Newman *et al.* (1998) tried to enhance plant metabolic abilities towards TCE (trichloroethylene), EDB (ethylene dibromide) by introduction of a gene coding for human cytochrome P450 IIE1 combined with oxidoreductase and cytochrome B5, and express the complex in microsomes of transformed poplar and tobacco.

Attempts to genetically improve the PCB-converting ability of plants are a further example of this approach in environmental remediation. While plants detoxify PCBs by hydroxylation and conjugation to hydrophilic molecules, followed by storage, some bacteria can open the biphenyl ring, thus allowing further steps leading to mineralisation. For this reason, as described in Borovka *et al.* (1998) and in Macek and Macková (1999),.a bacterial gene responsible for biphenyl ring opening was chosen for these attempts:

- gene C (*bphC*) of the bacterial biphenyl operon from *Pseudomonas* sp., characterised by Sylvestre *et al.* (Hein *et al.* 1998).

- cloning into Agrobacterium tumefaciens

- plant transformation

The obtained transgenic plants are now being tested for the activity of the target enzyme, 2,3-dihydroxybiphenyl 1,2-dioxygenase (Francova *et al.* 2001).

4.3. IMPROVEMENT OF HEAVY METAL UPTAKE

Plants have developed their own systems for binding of heavy metals based largely on the synthesis of phytochelatins, described by Grill *et al.* (1989). Heavy metal binding in plants is normally achieved, as reviewed by Kotrba *et al.* (1999) by phytochelatins and phytosiderophores (Rudolph *et al.* 1982). There are different attempts to improve the heavy metal accumulation capacity of high biomass yielding plants. The other option is to use traditional breeding to produce e.g. alpine pennycress plants that grow faster and taller.

In the last decade more attempts were described to prepare transgenic plants bearing genes for metal-binding proteins, metallothioneins (MTs) of different origin, human MTs, animal MTs and yeast MT genes (CUP1 gene, Truksa *et al.* 1997). Yancey *et al.* (1996) found increased Cd accumulation in tobacco plantlets bearing the CUP1 gene. Cadmium partitioning in transgenic tobacco plants expressing a mammalian metallothionein gene is described in detail by Dorlhac de Borne *et al.* (1998). The authors found increased Cd content in roots, but decreased in leafs.

Further enhancement of the performance of transgene products may be expected by implanting an additional heavy metal binding site, like a polyhistidine tail known for its high affinity to heavy metals. Such a gene encoding for 6 histidines was obtained from a commercial plasmid pTrc and cloned into *Agrobacterium tumefaciens* vector for delivery into plants (Macek *et al.* 1996). The obtained transgenic plants were tested, preliminary results showing two to three times higher Cd accumulation in tobacco plants bearing a construct with polyhistidine bound to yeast metallothionein in comparison with the control tobacco plants (Macek *et al.* 2000c). Already before transformation of plants the different prepared constructs were tested in *E. coli* in a series of growth curves, and together with improved resistance a substantial increase of Cd accumulation was found (Macek *et al.* 1997b, 1998).

Concerning the treatment of heavy metal contaminated fields, an interesting approach represents phytoremediation of mercury or methylmercury by genetically modified plants, using:

- merA - bacterial mercuric reductase (reducing mercury ions into volatile metal)

- merB - bacterial organomercurial lyase

These genes separately or both genes together were cloned into *Arabidopsis* or tobacco by Heaton *et al.* (1998) and proved to function well in soil. In this way a mercury-resistant transgenic plant was produced, that volatilises mercury into the atmosphere. Rugh *et al.* (1998) described transformation of poplar pro-embryogenic masses by microprojectile bombardment with modified *merA* constructs, obtaining plantlets releasing elemental mercury at high rate. The question arises, if volatilisation is an acceptable solution in this case, but the combined use of both enzymes appears to solve the immediate problem.

5. Conclusions

The main advantages of phytoremediation in comparison with classical remediation methods can be summarised (Salt *et al.* 1995, Schnoor *et al.* 1995):

- it is far less disruptive to the environment,

- there is no need for soil disposal sites,

- it has a high probability of public acceptance,

- avoids excavation and heavy traffic, and finally,

- it has potential versatility to treat a diverse range of hazardous materials.

Considering these factors, and the much lower cost expected for phytoremediation, it appears that it will allow for clean-up operations in much larger scale than possible by other methods. The process is relatively inexpensive, using the same equipment and supplies as agriculture.

Phytoremediation has a number of inherent technical limitations. The contaminant must be within, or drawn toward, the root zones of plants. This implies water, depth, nutrient, atmospheric, physical and chemical limitations. The site must be large enough to make farming techniques appropriate. There may be also a considerable delay in time needed for obtaining satisfactory clean-up results between phytoremediation and "dig and dump" techniques. In addition:

- formation of vegetation may be limited by extremes of environmental toxicity

- contaminants collected in leaves can be released again to the environment during litter fall

- contaminants can be accumulated in energy crops

- in some cases the solubility of some contaminants may be increased resulting in greater environmental damage and/or pollutant migration

The growing knowledge about the factors that are important in phytoremediation will provide a basis for genetic modification of plants directed to improved performance. We trust that molecular biology will allow to obtain plants tailored specifically for particular applications (Macek *et al.* 1998). These changes will include transforming of the plants to add specific proteins or peptides for binding and transporting xenobiotics, increasing the quantity and activity of plant biodegrading enzymes including those that are exported into the rhizosphere and the surrounding soil in order to improve the performance of soil bacteria. Plant-fungal interactions would also seem ripe for exploitation in this area, particularly mycorhizial associations. Alternatively, transgenic plants can be transformed to harbour microbial genes for biodegradation. This is already a routine practice in the engineering of many herbicide-resistant plants and their field testing, product development, and registration are well advanced. The concept could be extended to address additional xenobiotics.

The concept of manipulating plant genes that regulate toxic metal uptake or degradation of organic xenobiotics is a cutting-edge research. The likelihood of public acceptance and the fact that permits for field testing of genetically engineered plants now vastly outnumber permits for genetically altered microbes represent a good reason for moderate optimism concerning the use of phytoremediation.

Acknowledgement

Our work described in this chapter is sponsored by grant 203/99/1628 of the Grant Agency of the Czech Republic and research project of the DFG No: 436TSE113/35/0

References

Anonymous (1998) A citizens guide to phytoremediation. EPA Technology Fact Sheet, US EPA 542-F-98-011. US Environmental Protection Agency, Technology Innovation Office, August 1998, pp. 1-6

Baker A.J.M., McGrath S.P., Sidoli C.M.D. and Reeves R.D. (1994) The possibility of *in situ* heavy metal decontamination of polluted soils using crops of metal accumulating plants. Resources, Conservation and Recycling 11, 41-49

Batkhuugyin E., Rydlova J., Vosatka M. (2000) Effectiveness of indigenous and non-indigenous isolates of arbuscular mycorhizial fungi in soils from degraded ecosystems and man-made habitats. Appl Soil Ecol. 427, 1-11

Betts K.S. (1997) Native aquatic plants remove explosives. Environ. Sci.Technol. 31, 7, 304A

Bhadra R., Wayment D. G., Hughes J.B., Shanks J.V. (1999) Confirmation of conjugation processes during TNT metabolism by axenic plant tissues. Environ. Sci. Technol. 33, 446-452

Borovka R., Szekeres M., Macek T., Kotrba P., Sylvestre M. and Macková M. (1998) First steps in attempt to enhance the ability of plants to metabolise polychlorinated biphenyls by introduction of bacterial genes to plant DNA. Int. Biodeter. Biodegrad. 42, 243

Boyajian G.E. and Carreira L.H. (1997) Phytoremediation: A clean transition from laboratory to marketplace? Nature Biotechnol. 15, 127-128

Bringezu K., Lichtenberger O., Leopold I., Neumann D. (1999) Heavy Metal Tolerance of *Silene vulgaris*. J. Plant Physiol. 154, 536-546

Brown S.L., Chaney R.L., Angle J. S. and Baker A.J.M. (1995) Zinc and cadmium uptake by hyperaccumulator *Thlaspi caerulescens* and metal tolerant *Silene vulgaris* grown on sludge-amended soils. Environ Sci Technol. 29:10, 1581-1585

Burd G.I., Dixon D.G. and Glick B.R. (1998) A plant growth promoting bacterium that decreases nickel toxicity in seedlings. Appl. Environ. Microbiol. 64, 3663-3668

Burkhard J., Macková M., Macek T., Kucerová P., Demnerová K. (1997) Analytical procedure for PCB transformation by plant tissue cultures. Anal. Comm. Roy. Soc. 34, 10, 287-290

Butler J.M., Groeger A.W. and Fletcher J.S. (1992) Characterisation of monochlorinated biphenyl products formed by Paul's Scarlet Rose cells. Bull. Environ. Contam. Toxicol. 49, 821-826

Cole D. (1983) Oxidation of xenobiotics in plant. Progress in Pest. Biochem. Toxicol. 3, 199-253

Cunningham S.D., Berti W.R. and Huang J.W. (1995) Phytoremediation of contaminated soils. TIBTECH 13, 393-397

Cunningham S.D. and Ow D.W. (1996) Promises and prospects of phytoremediation. Plant Physiol. 110, 715-719

Dannel F., Floesser-Mueller H., Schwack W. (1995) Aufnahme von Tetrachlorethylen über das Wurzelsystem von Tomaten. BIOforum 18, 207-210

Dixon P.D., Cummins I., Cole D.J. and Edwards R. (1998) Glutathione-mediated detoxification systems in plants. Curr Opinion Plant Biol, 1, 258-266

Donelly P.K., Hedge R.S. and Fletcher J.S. (1994) Growth of PCB-degrading bacteria on compounds from photosynthetic plants. Chemosphere 28, 984-988)

Donnelly P.K. and Fletcher J.S. (1994) Potential use of mycorrhizal fungi as bioremediation agents. In: Anderson T.A., [Ed.], Bioremediation through rhizosphere technology, ACS Symposium Series No. 563, American Chemical Society, pp. 93-99

Doran P. M. (1997) Hairy Roots: Culture and Applications. Harwood Academic Publishers, London

Dorlhac de Borne F.D., Elmayan T., de Roton Ch., de Hys L., Tepfer M. (1998) Cadmium partitioning in transgenic tobacco plants expressing a mammalian metallothionein gene. Molecular Breeding 4, 83- 90

Duc R., Vanek T., Soudek P., Schwitzguebel J.P. (1999a) Accumulation and transformation of sulphonated aromatic compounds by rhubarb (*Rheum palmatum*) cells. Int. J. Phytoremediation, 1, 3, 255-271

Duc R., Vanek T., Soudek P., Schwitzguebel J.P. (1999b) Experiments with rhubarb in Europe. Soil and Groundwater Cleanup, 2/3, 27-30

Flathman P.E., Lanza G.R. and Glass D.J. (1999) Phytoremediation Issue. Soil and Groundwater Cleanup 2, 4-11

Fletcher J.S., Groeger A.W, and McFarlane J.C. (1987) Metabolism of 2-chlorobiphenyl by suspension cultures of Paul's Scarlet Rose. Bull. Environ. Contam. Toxicol. 39, 960-965

Fletcher J.S., Donnelly P.K. and Hedge R.S. (1995) Biostimulation of PCB-degrading bacteria by compounds released from plant roots. In: Hinchee R.E., Anderson D.B. and Hoeppel R.E. (Eds.), Bioremediation of recalcitrant organics, Battelle Press, Columbus, pp. 131-136

Forman S., Kučerová P., Macková M., Macek T., and Ruml T. (1998) Toxicity of PCB evaluated by cell culture systems. Int. Biodeterior. Biodegrad. 42, 250

Francová K., Macek T., Demnerova K., and Macková M., (2001) Transgenic plants – A potential tool for decontamination of environmental pollutants. Chem. Listy 95, 630-637

French C.E., Rosser S.J., Davies G.J., Nicklin S. and Bruce N. (1999) Biodegradation of explosives by transgenic plants expressing pentaerythritol tetranitrate reductase. Nature Biotechnol, 17, 491-494

Gilbert E.S and Crowley D.E (1997) Plant compounds that induce polychlorinated biphenyl biodegradation by *Arthrobacter sp.* strain B1B. Appl. Environ. Microbiol. 63, 1933-1938

Grill E., Loffler S., Winnacker E-L. and Zenk M.H. (1989) Phytochelatins, the heavy-metal-binding peptides of plants, are synthesised from glutathione by a specific γ-glutamylcysteine dipeptidyl transpeptidase (phytochelatin synthase). Proc. Nat. Acad. Sci. USA, 84, 6838-6846

Groeger A.W. and Fletcher J.S. (1988) The influence of increasing chlorine content on the accumulation and metabolism of polychlorinated biphenyls by Paul's Scarlet Rose cells. Plant Cell Reports, 7, 329-332

Hammer D.A. (1989) Constructed wetlands for wastewater treatment. Municipal, Industrial and Agricultural. Chelsea. Mich. Lewis Publishers

Harms H. and Kottutz E. (1990) In: Progress in Plant Cellular and Molecular Biology, Nijkamp H.J.J., van der Plas L.H.W. and van Aartrijk J., [Eds.], Kluwer Academic Publishers, Dordrecht, pp. 650-655

Heaton A.C.P., Rugh C.L., Wang N.-J., Meagher R.B. (1998) Phytoremediation of Hg-polluted soils by genetically engineered plants. J. Soil Cont., 7:4, 497-509

Hedge R.S. and Fletcher J.S. (1996) Influence of plant growth stage and season on the release of root phenolics by mulberry as related to development of phytoremediation technology. Chemosphere 32, 2471-2479

Hein P., Powlowski J., Barriault D., Hurtubise Y., Ahmad D. and Sylvestre M. (1998) Biphenyl-associated meta-cleavage dioxygenases from *Comamonas testosteroni* B-356. Can. J. Microbiol. 44, 42-49

Hubálek M., Vanek T. (1998) Degradation of 2,4,6-trinitrotoluene by cell suspension of *Solanum aviculare*. Int. J. Biodeterior. Biodegr. 42, 251

Hughes J.B., Shanks J., Vanderford M., Lauritzen J. and Bhadra R. (1997) Transformation of TNT by aquatic plants and plant tissue cultures. Environ. Sci. Technol. 31, 266-271

Jones R.L. (1994) ASPP recommends hazardous waste remediation technologies to DOE. ASPP Newsletter, 21, 6, 12-13

Káš J., Burkhard J., Demnerová K., Košťál J., Macek T., Macková M. and Pazlarová J. (1997) Perspectives in biodegradation of alkanes and PCBs. Pure Appl Chem 69, 2257-2369

Kayser A., Felix H.R. (1998) Five years of phytoremediation in the field. In: Book of Ext. Abstr. Int. Workshop - Innovative Potential of Advanced Biological Systems for Remediation. (K.N. Timmis, Ed.), Technical University Hamburg-Harburg, 2-4 March 1998, pp. 81-86

Komosa D., Gennity I., Sandermann H. (1992) Plant metabolism of herbicides with C-P bonds: Glyphosate. Pest. Biochem. Physiol. 43, 85-94

Komosa D., Sandermann H. (1992) Plant metabolism of herbicides with C-P bonds: Phosphinotricin. Pest. Biochem. Physiol. 43, 95-102

Kotrba P., Macek T. and Ruml T. (1999) Heavy-metal binding peptides and proteins in plants. A review. Coll. Czech Chem. Commun. 64, 1057-1086

Krijger G.C., Harms A.V., Leen R., Verburg T.G., Wolterbeek B. (1999) Chemical forms of technetium in tomato plants; TcO₄⁻, Tc-cysteine, Tc- glutathione and Tc-proteins. Environmental and Experimental Botany 42, 69-81

Kucerová P., Poláchová L., Macek T., Burkhard J., Tříska J., and Macková M. (1998a) Use of plant tissue cultures for model studies to evaluate the ability of plants to metabolise PCB. Int. Biodeterior. Biodegrad. 42, 249

Kucerová P., Poláchová L., Macek T., Burkhard J., Pazlarová J., Demnerová K. and Macková M (1998b) Transformation of PCBs by plant cell cultures and relation to the production of plant peroxidases. Int. Biodeterior. Biodegrad. 42, 250

Kucerová P., Macková M., Poláchová L., Burkhard J., Demnerová K., Pazlarová J. and Macek T. (1999) Correlation of PCB transformation by plant tissue cultures with their morphology and peroxidase activity changes. Coll. Czech Chem. Commun. 64 (9), 1497-1509

Kucerová P., Macková M., Chromá L., Burkhard J., Triska J., Demnerová K. and Macek T. (2000) Metabolism of polychlorinated biphenyls by *Solanum nigrum* hairy root clone SNC-9O and analysis of transformation products. Plant Soil, 225, 109-115

Lamoureux G.L. and Frear D.S. (1979) Pesticide metabolism in higher plants: In vitro enzyme studies. In: Paulson G.D., Frear D.S. and Marks E.P. [Eds.], Xenobiotic Metabolism. In vitro Methods, American Chemical Society Symposium Series. Vol 97, ACS, Washington DC, pp. 72-128.

Langebartels C. and Harms H. (1984) Metabolism of pentachlorophenol in cell suspension cultures of soybean and wheat: Pentachlorophenol glucoside formation. Z. Pflanzenphysiol. 113, 201-211

Lee I. and Fletcher J.S. (1992) Involvement of mixed function oxidase systems in PCB metabolism by plant cells. Plant Cell Reports 11, 97-100

Macek T. (1989). Poroporo, *Solanum aviculare, S. laciniatum*: In vitro culture and the production of solasodine, In: Bajaj Y.P.S. [Ed.], Biotechnology in Agriculture and Forestry, 7, Springer Verlag Heidelberg, pp. 443-467

Macek T., Kotrba P., Suchová M., Skácel F, Demnerová K., Ruml T. (1994) Accumulation of cadmium by hairy root cultures. Biotechnol. Lett., 16, 6, 621-624

Macek T., Macková M., Truksa M., Singh-Cundy A., Kotrba P., Yancey N., Scouten W.H. (1996) Preparation of transgenic tobacco with a yeast metallothionein combined with a polyhistidine tail. Chem. Listy, 90, 690

Macek T., Kotrba P., Ruml T., Skácel F. and Macková M. (1997a) Accumulation of cadmium by hairy root cultures. In: P.M. Doran (ed.), Hairy Roots: Culture and Application, Harwood Academic Publishers, London, pp. 133-138

Macek T., Macková M., Kotrba P., Truksa M., Singh-Cundy A., Scouten W.H. and Yancey N. (1997b) Attempts to prepare transgenic tobacco with higher capacity to accumulate heavy metals containing yeast metallothionein combined with a polyhistidine, In H. Verachtert and W. Verstraete (eds.), Environmental Biotechnology, Proc. Int. Symp., Oostende, April 1997, Technological Institute Antwerp, pp. 263-266

Macek T., Macková M., Burkhard J., Demnerová K. (1998) Introduction of green plants for the control of metals and organics in remediation. In: Effluents from Alternative Demilitarisation Technologies. (F.W. Holm, Ed), NATO PS Ser. 1, Vol. 12, Kluwer Acad. Publishers, Dordrecht, pp. 71-85

Macek T. and Macková M. (1998) Phytoremediation – use of plants for control of organics and metals in the environment. Chemical Papers 52, 581-582

Macek T. and Macková M. (1999) Phytoremediation – the use of plants to remove xenobiotics and pollutants from the environment, including transgenic plants tailored for this purpose. Biologia 54, 70-73

Macek T., Macková M., Kucerová P., Burkhard J., Kotrba P., Demnerová K. (1999) Phytoremediation – its possible application in chemical weapons demilitarisation. In: Proc. Int. Congr. Chemical Weapons Demilitarisation, CWD99 Vienna, July 1999, (I. Chillcott, Ed.), DERA UK, pp. 865-912

Macek, T., Macková, M., Káš, J. (2000a) Exploitation of plants for the removal of organics in environmental remediation. Biotechnol. Advances 18 (1), 23-35

Macek T., Macková M., Kučerová P., Poláchová L., Burkhard J. and Demnerová K. (2000b) Phytoremediation – illustrated by plant conversion of PCBs. Proceedings of the 9th European Congress of Biotechnology, (M. Hofman, ed.), CD-ROM, Branche Belge de la Société de Chimie Industrielle, ISBN 805215-1-5

Macek T., Mackova M., Pavlikova D., Szakova J., Truksa M., Singh-Cundy A., Kotrba P., Yancey N., Scouten W.H. (2000c) Cadmium uptake by transgenic tobacco clones. Chem Listy 94 (8), 745-746

Macková M., Macek T., Očenášková J., Burkhard J., Demnerová K. and Pazlarová J. (1996) Selection of the potential plant degraders of PCBs. Chem Listy 90 (9), 712-713

Macková M., Macek T., Burkhard J., Ocenasková J., Demnerová K., Pazlarová J. (1997a) Biodegradation of polychlorinated biphenyls by plant cells. Int. Biodeter. Biodegrad., 39:4, 317-325

Macková M., Macek T., Kucerová P., Burkhard J., Pazlarová J. and Demnerová K. (1997b) Degradation of polychlorinated biphenyls by hairy root culture of *Solanum nigrum*. Biotechnol. Lett., 19, 8, 787-790

Macková M., Macek T., Kucerová P., Burkhard J., Triska J., Demnerová K. (1998) Plant tissue cultures in model studies of transformation of polychlorinated biphenyls. Chemical Papers 52, 599-600

Macková M., Macek T., Kucerová P., Poláchová L., Burkhard J., Triska J. (1999) The beneficial effect of plants on detoxification of environmental pollutants. Biologia 54, 76-77

Macková M., Kucerova P., Demnerova K., Leigh M.B., Polachova L., Totevova S., Burkhard J., Kastanek F. and Macek T. (2000a) Practical use of bioremediation for PCB removal from contaminated soil. In: Proceedings of the 9th European Congress on Biotechnology, Brussels, July 1999 (M. Hofman, ed.), CD-ROM, Branche Belge de la Société de Chimie Industrielle, ISBN 805215-1-5

Macková M, Ferri E.N., Demnerová K. and Macek T. (2001a) Quantitative chemiluminiscent detection of plant peroxidases using commercial kit designed for blotting assays. Chem Listy, 95, 130-132

McCully M.E. (1999) Roots in soil: Unearthing the Complexities of Roots and Their Rhizospheres. Plant Phystol. Plant Mol. Biol. 50, 695-718

Morel J.L., Chaineau C.H., Schiavon M. and Lichtfouse E. (1999) The role of plants in the remediation of contaminated soils. In: Bioavailability of Organic Xenobiotics in the Environment, Ph. Baveye *et al.* (eds.), Kluwer Academic Publishers, Dordrecht, 429-449

Moza P., Weisgerber I., Klein W. and Korte F. (1974) Metabolism of 2,2´-dichlorobiphenyl-^{14}C in two-plant-water-soil-system. Bull. Environm. Contam. Toxicol. 12, 541-546

Moza P., Weisgerber I., and Klein W. (1976) Fate of 2,2´-dichlorobiphenyl-^{14}C in carrots, sugar beets, and soil under outdoor conditions. J. Agric. Food Chem. 24, 881-885

Nepovim A., Vanek T. (2000) Alternative ways of 2,4,6-TNT degradation by cell-free extract from tissue culture of *Solanum aviculare*. In: Proc. Conf. Phytoremediation 2000: State of the art in Europe (an international comparison) 6-8. 4. 2000, Greece, p. 36

Newman L.A., Doty S.L., Gery K.L., Heilman P.E., Muiznieks I., Shang T.Q., Siemieniec S.T., Stran S.E., Wang X.-P., Wilson A.M., Gordon M.P. (1998) Phytoremediation research at the University of Washington. J. Soil Cont. 7, 4, 531-542

O´Sullivan A.D., McCabe O.M., Murray D.A., Otte M.L. (1999) Wetlands for rehabilitation of metal mine wastes. Biol. Environ. 99B, 11-17

Olson P.E. and Fletcher J.S. (1999) Field evaluation of mulberry root structure with regard to phytoremediation. Bioremediation J, 3 (1), 1-7

Olson P.E. and Fletcher J.S. (2000a) Ecological recovery of vegetation at a former industrial sludge basin and its implications to phytoremediation. In Press. Environmental Science and Pollution Research 7, in press.

Olson P.E. and Fletcher J.S. (2000b) Natural attenuation / phytoremediation in the vadose zone of a former industrial sludge basin. Environmental Science and Pollution Research, submitted.

Pletsch M., Santos de Araujo B., Charlwood B.V. (1999) Novel biotechnological approaches in environmental remediation research. Biotechnology Advances 17, 679-687

Plewa M.J., Wagner E.D. (1992) Metabolic activation of promutagens into mutagenic compounds by plants. Ann Rev. Genet. 27, 93-102

Pradhan S.P., Conrad J.R., Paterek J.R. and Srivastava V.J. (1998) Potential of phytoremediation for treatment of PAHs in soil at MGP Sites. J. Soil Contam. 7, 467-480

Rout G.R., Samantaray S., Das P. (1999) In vitro selection and biochemical characterisation of zinc and manganese adapted callus lines in *Brassica* spp. Plant Science 137, 89-100

Rudolph A., Becker R., Scholz G., Procházka Z, Toman J., Macek T. and Herout V. (1985) The occurrence of the amino acid nicotianamine in plants and microorganisms. A reinvestigation. Biochem. Physiol. Pflanzen, 180, 557-563

Rugh C.L., Senecoff J.F., Meagher R.B., Merkle S.A. (1998) Development of transgenic yellow poplar for mercury phytoremediation. Nature Biotechnology 16, 925-928

Salt D.E., Blaylock M., Kumar N. P.B.A., Dushenkov V., Ensley B.D., Chet I. and Raskin I. (1995) Phytoremediation: A novel strategy for the removal of toxic metals from the environment using plants. Bio/Technology, 13, 5, 468-474

Sandermann H. (1992) Plant metabolism of xenobiotics. Trends Biochem. Sci. 17, 82-84.

Sandermann H. (1994) Higher plant metab. of xenobiotics: the green liver concept. Pharmacogenetics 4, 225-241

Sanita di Toppi L., Gabbrielli R. (1999) Response to cadmium in higher plants. Environmental and Experimental Botany 41, 105-130

Schnoor J.L., Licht L.A., McCutcheon S.C., Wolfe N.L., Carreira L.H. (1995) Phytoremediation of organic contaminants. Environ. Sci. Technol. 29, 318-323

Shanks J.V. and Morgan J. (1999) Plant "hairy root" culture. Current Opinion in Biotechnology 10, 151-155

Soudek P., Podlipná R., Lipavská H., Vanek T. (1998a) Bioaccumulation of heavy metals by hairy root culture of Armoracia rusticana. Pharmaceut. Pharmacol. Letters 8, 2

Soudek P., Podlipná R., Vanek T. (1998b) Phytoremediation of heavy metals by hairy root culture of Armoracia rusticana. Int. J. Biodeterior. Biodegrad. 42, 235-236

Stiborová M., Anzenbacher P. (1991) What are the principal enzymes oxidizing the xenobiotics in plants: cytochromes P450 or peroxidases? (A hypothesis). Gen. Physiol. Biophys. 10, 209-216

Susarla S., Bacchus T.S., Wolfe N.L. and McCutcheon C.S (1999) Phytotransformation of perchlorate using parrot feather. Soil and Groundwater Cleanup, 2, 20-23

Tebbe C.C., Schwieger F., Munch J.C., Puehler A., Keller M. (1998) Field release of genetically engineered bioluminescent Sinorhizobium meliloti strains. In: Sustainable Agriculture for Food, Energy and Industry, N. El Bassam, R.K. Behl, B. Prochnow, eds., James and James, Science Publishers, London, pp.450-452

Tlustos P., Balik J., Pavlíková D., Száková J. (1997) The uptake of cadmium, zinc, arsenic and lead by chosen crops. Rostl. Výr., 43, 10, 487-494

Truksa M., Singh-Cundy A., Macek T., Kotrba P., Macková M., Yancey N., Scouten W.H. (1996) Transgenic plants expressing metal binding proteins in phytoremediation. Chem. Listy, 90, 9, 707

Vaněk T., Urmantseva V.V., Wimmer Z., Macek T. (1989) Biotransformation of 2-(4-methoxybenzyl)-1-cyclohexanone by Dioscorea deltoidea free and immobilized plant cells. Biotechnol. Lett., 11, 234-248

Vaněk T., Wimmer Z., Macek T., Šaman D., Svatoš A. and Romanuk M. (1989) Stereochemistry of the enzymatic reduction of 2,4-methoxybenzyl-1-cyclohexanone by Solanum aviculare cells in vitro. Biocatalysis, 2, 265-272

Vymazal J., Brix H., Cooper P.F., Haberl R., Perfler R., Laber J. (1998) Removal mechanisms and types of constructed wetlands. In: Constructed Wetlands for Wastewater Treatment in Europe (Vymazal J., Brix H., Cooper P.F., Green M.B. and Haberl R., Eds.), Backhuys Publishers, Leiden

Wilken A., Bock C., Bokern M. and Harms H. (1995) Metabolism of different PCB congeners in plant cell cultures. Environ. Toxicol. Chem. 14, 2017-2022

Wimmer Z., Macek T., Vaněk T., Streinz L. and Romaňuk M. (1987) Biotransformation of 2-(4-methoxybenzyl)-1-cyclohexanone by cell cultures of Solanum aviculare. Biol. Plant., 29, 88-93

Wollgiehn R., Neumann D. (1999) Metal Stress Response and Tolerance of Cultured Cells from Silene vulgaris and Lycopersicon peruvianum: Role of Heat Stress Proteins. J. Plant Physiol. 154, 547-553

Wright D.J. and Otte M.L. (1999) Wetland plants effects on the biogeochemistry of metals beyond the rhizosphere. Biol. Environ. 99B, 3-10

Yancey N., McLean J.E., Sims R.C., Scouten W.H, Singh-Cundy A., Kotrba P., Macková M., Macek T., and Truksa M. (1996) Cadmium accumulation in transgenic and non-transgenic tobacco plants. In: Book of Abstracts of the HSRC/WERC Joint Conference on the Environment, Albuquerque, New Mexico, May, (1996)

Yateem A., Balba M.T., El-Nawawy A.S. and Al-Awadhi N. (1999) Experiments in phytoremediation of Gulf War contaminated soil. Soil and Groundwater Cleanup 2, 31-33

Zakharova E.A., Kosterin P.V., Brudnik V.V., Sherbakov .A.A., Ponomarjov A.S., Ignatov V.V. (2000) Proceedings of the 9th European Congress of Biotechnology ECB9, CD-ROM (M. Hofman, Ed.), Branche Belge de la Société de Chemie Industrielle, Brussels, ISBN 805215-1-5

Zechendorf B. (1999) Sustainable development: how can biotechnology contribute? TIBTECH, 17, 219-225

INDEX

¹⁴C.............................. 94; 96; 97; 98; 100; 101; 104; 107; 108; 110; 111; 136
2,4-D 97; 99; 105; 106; 107; 122
active plume management...73; 79
Adamsite 127
aerobic composting 103; 104; 105; 106; 107; 109; 110; 111
aerobic in situ bioremediation.. 79
artificial wetland 115; 118; 121
benzene 8; 12; 75; 76; 80; 81; 83; 86; 119
bioavailability..vi; 9; 16; 37; 38; 93; 98; 99
biological degradation...9; 10; 14; 47; 73; 90
biological soil remediation... 17; 18; 73
biological treatment...vi; 7; 12; 21; 37; 46; 69
bioprocessing 51; 59
bioremediation ...v; vi; 10; 11; 18; 19; 23; 29; 30; 31; 37; 54; 69; 70; 73; 74; 76; 81; 86;
    89; 103; 104; 105; 115; 118; 127; 134; 136
bound residue 94; 98; 100
BTEX......................... 8; 12; 52; 56; 74; 76; 79; 81; 83; 84; 85; 86; 87; 90
BTEX removal 81
cell suspension cultures.. 124; 125; 135
chemical analysis 18
chemical war agent.. 127
chemical weapons demilitarisation 116; 135
CIASR method.. 25
Clark... 127
Clark1... 127
Clark2... 127
Comprehensive impact assessment of site remediation 25; 33
contaminated soil ..v; 7; 8; 13; 16; 25; 27; 30; 32; 37; 38; 46; 47; 51; 53; 54; 58; 62; 66;
    69; 70; 81; 89; 103; 104; 106; 111; 118; 119; 121; 122; 123; 128; 129; 133; 136; 137
CW agent 127
CWD 116; 121
CWs... 127
cyclotrimethylenetrinitramine 121
cyprodinil 97; 100
dechlorination...77; 78; 83; 85; 86; 89
degradability... 9; 11
detoxification .... vi; 93; 95; 99; 106; 107; 111; 117; 121; 127; 129; 133; 136
diphenylaminechloroarsine 127
diphenylchloroarsine 127
economics... 52; 65

EDC.. 116
endocrine-disrupting chemical 116
environmental balancing 23; 25; 26; 30; 31; 32
environmental merit index 28
enzymes...87; 93; 94; 95; 96; 98; 99; 100; 101; 115; 117; 118; 119; 120; 121; 125; 128; 131; 132; 137
enzymology v
ethylbenzene... 8; 12; 75; 119
ex situ processes... 19; 55
exudates... 117; 118; 120
fences 81; 82
field experiments... 129
fine grained soil.. 21; 37
genetic engineering 115; 118; 124; 128; 129; 130
hairy root cultures 124; 125; 135
harbour site... 81
HCH removal 80
heavy metal removal 122
heavy metals.............. vi; 12; 66; 115; 116; 119; 122; 124; 125; 126; 127; 131; 135; 137
humification ....vi; 8; 19; 95; 103; 106; 107; 109; 110
humus... 93; 94; 98; 99; 100; 110
in situ.... v; 10; 13; 16; 18; 19; 20; 30; 54; 68; 70; 73; 74; 75; 77; 79; 81; 83; 85; 86; 87; 88; 89; 116; 133
in situ methods 19
indicators for in situ biodegradation... 84
laboratory facility.. 39; 42
laccase 99; 100; 118
life cycle assessment .... v; 23; 24; 25; 26; 27; 28; 30; 31; 32
metallothionein.. 131; 134; 135
monitored natural attenuation... 19; 86
MTs.. 131
natural attenuation........................ v; 17; 19; 73; 74; 79; 81; 82; 83; 85; 86; 87; 88; 119
natural attenuation rate 85
nitroaromatics............................... 103; 105; 106; 107; 109; 110
NMR spectroscopy... 96; 97; 100; 111
NMVOC... 26
non methane volatile organic compound...................................... 26
non-extractable ¹⁵N-TNT residues... 109
oxidative coupling... 93; 96; 100
PAH......8; 12; 15; 16; 37; 40; 41; 43; 44; 45; 46; 47; 48; 52; 62; 63; 64; 68; 79; 88; 122
PAH-degradation 44; 45; 46; 47; 63
PCB............ 8; 12; 52; 115; 118; 122; 125; 127; 128; 129; 130; 133; 134; 135; 136; 137
PCB transformation products 128
PCDD... 8; 12
PCDF.. 8; 12
physico-chemical conditions 10

phytodegradation... 115; 118; 122
phytoextraction..................................... 115; 118; 119; 122; 123; 124
phytoremediation . vi; 19; 52; 66; 68; 116; 117; 118; 120; 121; 122; 123; 124; 125; 126; 130; 131; 132; 133; 134; 136; 137
phytovolatilisation.. 115; 122
pilot plant 19; 39; 40; 41; 42; 43; 44; 45; 47; 48; 61; 62; 65; 68; 69
planning...v; 23; 24; 30; 32; 33; 47; 67
plant in vitro culture 124
plume behaviour.. 83; 85
polychlorinated biphenyl...........................8; 12; 127; 133; 134; 135; 136
polychlorinated dibenzodioxin 8; 12
polychlorinated dibenzofuran.. 8
polycyclic aromatic hydrocarbon 8; 12; 37; 48; 74
process-engineering characterization 41; 43
Rademarkt, Groningen 77
RDX 121; 122
reactor process.. v; 68
REC.........................23; 24; 25; 27; 28; 29; 30; 31; 32; 33
recycling...v; 51; 52; 53; 54; 62; 63; 68; 127
remediation techniques... 73; 75; 77
rhizofiltration 115; 118; 122; 124
rhizoremediation 116; 118; 129
Sapromat system 40
sediment 48; 55; 62; 63; 69; 83
slurry decontamination process 51
soil remediation.......................v; 23; 24; 25; 26; 27; 30; 31; 32; 33; 68
solid waste...20; 51; 52; 53; 55; 61; 68
source zone 73; 75; 77; 84; 89
suspension reactor 21; 37; 39; 45; 47; 48
sustainability 32; 87
TCE.......................76; 77; 78; 83; 89; 90; 120; 122; 130
TNT-contaminated soil 103; 104; 105; 106; 110
toluene.......................................8; 12; 41; 75; 76; 89; 119
total petroleum hydrocarbons... 8
TPH.. 8; 76
transformation of TNT 19; 103; 106
trichloroethylene 120
VOCH 8; 12
volatile halogenated hydrocarbons.. 8
xenobiotics removal 122
xylene.. 8; 119

Printed in the United States
900900001B